Solutions Manual *for the* Chemical Engineering Reference Manual

Fifth Edition

Randall N. Robinson, PE

Professional Publications, Inc. • Belmont, CA

In the ENGINEERING LICENSING EXAM AND REFERENCE SERIES

Engineer-In-Training Reference Manual
EIT Review Manual
Engineering Fundamentals Quick Reference Cards
Engineer-In-Training Sample Examinations
Mini-Exams for the E-I-T Exam
1001 Solved Engineering Fundamentals Problems
Fundamentals of Engineering Exam Study Guide
Diagnostic F.E. Exam for the Macintosh
Fundamentals of Engineering Video Series:
 Thermodynamics
Civil Engineering Reference Manual
Quick Reference for the
 Civil Engineering PE Exam
Civil Engineering Sample Examination
Civil Engineering Review Course on Cassettes
101 Solved Civil Engineering Problems
Seismic Design of Building Structures
Seismic Design Fast
345 Solved Seismic Design Problems
Timber Design for the Civil P.E. Examination
246 Solved Structural Engineering Problems
Mechanical Engineering Reference Manual
Mechanical Engineering Quick Reference Cards
Mechanical Engineering Sample Examination
101 Solved Mechanical Engineering Problems
Mechanical Engineering Review Course
 on Cassettes
Consolidated Gas Dynamics Tables
Fire and Explosion Protection Systems

Electrical Engineering Reference Manual
Electrical Engineering Quick Reference Cards
Electrical Engineering Sample Examination
Chemical Engineering Reference Manual
 for the PE Exam
Quick Reference for the
 Chemical Engineering PE Exam
Chemical Engineering Practice Exam Set
Land Surveyor Reference Manual
Land Surveyor-In-Training Sample Examination
1001 Solved Surveying Fundamentals Problems

In the GENERAL ENGINEERING and CAREER ADVANCEMENT SERIES

How to Become a Professional Engineer
Getting Started as a Consulting Engineer
The Expert Witness Handbook:
 A Guide for Engineers
Engineering Your Job Search
Engineering Your Start-Up
Engineering Your Writing Success
Intellectual Property Protection:
 A Guide for Engineers
High-Technology Degree Alternatives
Metric in Minutes
Engineering Economic Analysis
Engineering Law, Design Liability, and
 Professional Ethics
Engineering Unit Conversions

SOLUTIONS MANUAL for the
CHEMICAL ENGINEERING REFERENCE MANUAL
Fifth Edition

Copyright © 1996 by Professional Publications, Inc. All rights reserved. No part of this publication may be reproduced, stored in a retrieval system, or transmitted, in any form or by any means, electronic, mechanical, photocopying, recording, or otherwise, without the prior written permission of the publisher.
Printed in the United States of America
Professional Publications, Inc.
1250 Fifth Avenue, Belmont, CA 94002
(415) 593-9119
Current printing of this edition: 1

Library of Congress Cataloging-in-Publication Data
Robinson, Randall N.
 Solutions manual for the chemical engineering reference manual, fifth edition / Randall N. Robinson.
   p. cm.
  ISBN 0-912045-93-0 (perfect bnd.)
  1. Chemical engineering--Examinations, questions, etc.
 I. Robinson, Randall N. Chemical engineering reference manual. 5th ed. II. Title
TP168.R645 1996
660'.076--dc20   96-16477
                                                          CIP

TABLE OF CONTENTS

1 MATHEMATICS 1

2 STOICHIOMETRY, HEAT, AND MATERIAL BALANCES 5

3 ENGINEERING ECONOMICS 13

4 THERMODYNAMICS 17

5 FLUID STATICS AND DYNAMICS 25

6 HEAT TRANSFER: CONDUCTION AND RADIATION 33

7 HEAT TRANSFER: CONVECTION AND EQUIPMENT 37

8 VAPOR-LIQUID PROCESSES 43

9 DISTILLATION, EVAPORATION, AND HUMIDIFICATION 47

10 LIQUID-LIQUID AND SOLID-LIQUID PROCESSES 53

11 KINETICS 59

The National Society of Professional Engineers

Whether you design water works, consumer goods, or aerospace vehicles; whether you work in private industry, for the U.S. government, or for the public; and whether your efforts are theoretical or practical, you (as an engineer) have a significant responsibility.

Engineers of all types perform exciting and rewarding work, often stretching new technologies to their limits. But those limits are often incomprehensible to nonengineers. As the ambient level of technical sophistication increases, the public has come to depend increasingly and unhesitatingly more on engineers. That is where professional licensing and the National Society of Professional Engineers (NSPE) become important.

NSPE, the leading organization for licensed engineering professionals, is dedicated to serving the engineering profession by supporting activities, such as continuing educational programs for its members, lobbying and legislative efforts on local and national levels, and the promotion of guidelines for ethical service. From local, community-based projects to encourage top-scoring high school students to choose engineering as a career, to hard-hitting lobbying efforts in the nation's capital to satisfy the needs of all engineers, NSPE is committed to you and your profession.

Engineering licensing is a two-way street: it benefits you while it benefits the public and the profession. For you, licensing offers a variety of benefits, ranging from peer recognition to greater advancement and career opportunities. For the profession, licensing establishes a common credential by which all engineers can be compared. For the public, a professional engineering license is an assurance of a recognizable standard of competence.

NSPE has always been a strong advocate of engineering licensing and supporter of the profession. Professional Publications hopes you will consider membership in NSPE as the next logical step in your career advancement. For more information regarding membership, write to the National Society of Professional Engineers, Information Center, 1420 King Street, Alexandria, VA 22314, or call (703) 684-2800.

PROFESSIONAL PUBLICATIONS, INC. • Belmont, CA

Notice to Examinees

Do not copy, memorize, or distribute problems from the Principles and Practice of Engineering (P&P) Examination. These acts are considered to be exam subversion.

The P&P examination is copyrighted by the National Council of Examiners for Engineering and Surveying. Copying and reproducing P&P exam problems for commercial purposes is a violation of federal copyright law. Reporting examination problems to other examinees invalidates the examination process and threatens the health and welfare of the public.

PROFESSIONAL PUBLICATIONS, INC. • Belmont, CA

SOLUTIONS FOR CHAPTER 1
MATHEMATICS

1.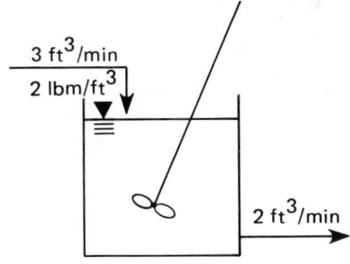

$$\text{input} - \text{output} = \text{accumulation}$$
$$Q = \text{flow, ft}^3/\text{min}$$
$$c_o = \text{concentration out}$$
$$t = \text{time, min}$$
$$V = \text{volume, ft}^3$$
$$c_i = \text{concentration in}$$

$$Q_i c_i - Q_o c_o = \frac{d}{dt}(Vc_o) = c_o \frac{dV}{dt} + V \frac{dc_o}{dt}$$

$$\frac{dV}{dt} = 1 \text{ ft}^3/\text{min}$$

$$\int dV = 1 \int dt$$

or $\quad V = t + k$

At $t = 0$, $V = 20$ ft^3, $k = 20$ ft^3, and $V = t + 20$.

$$(3)(2) - 2c_o = c_o(1) + (1t + 20)\frac{dc_o}{dt}$$

$$\int \frac{dc_o}{(6 - 3c_o)} = \int \frac{dt}{t + 20}$$

$$-\tfrac{1}{3}\ln(2 - c_o) = \ln(20 + t) + I$$

At $t = 0$, $c_o = 0$.

$$I = \ln\left(\frac{1}{20\sqrt[3]{6}}\right)$$

Substitute for I and rearrange.

$$c_o = 2 - \frac{2}{(1 + 0.05t)^3}$$

When $V = 30$, $t = 10$.

$$c_o = 2 - \frac{2}{[1 + (0.05)(10)]^3} = 1.407 \text{ lbm/ft}^3$$

When $V = 50$, $t = 30$.

$$c_o = 2 - \frac{2}{[1 + (0.05)(30)]^3} = 1.872 \text{ lbm/ft}^3$$

$1.872 < (2)(0.99) = 1.98$

The differential equation for the overflow is

$$(3)(2) - 3c_o = V\frac{dc_o}{dt}$$
$$V = 50$$

$$\int \frac{dc_o}{(6 - 3c_o)} = \frac{1}{50}\int dt$$
$$-\tfrac{1}{3}\ln(6 - 3c_o) = \frac{1}{50}t + I$$

At $t = 0$, $c_o = 1.872$.

$$I = \ln\left(\frac{1}{\sqrt[3]{0.384}}\right)$$

$$t = \frac{50}{3}\ln\left[\frac{0.384}{(6 - 3c_o)}\right]$$

When $c_o = 1.98$,

$$t = \frac{50}{3}\ln\left[\frac{0.384}{6 - (3)(1.98)}\right]$$
$$= 30.9 \text{ min}$$

total time $= 30 + 30.9 = \boxed{60.9 \text{ min}}$

2.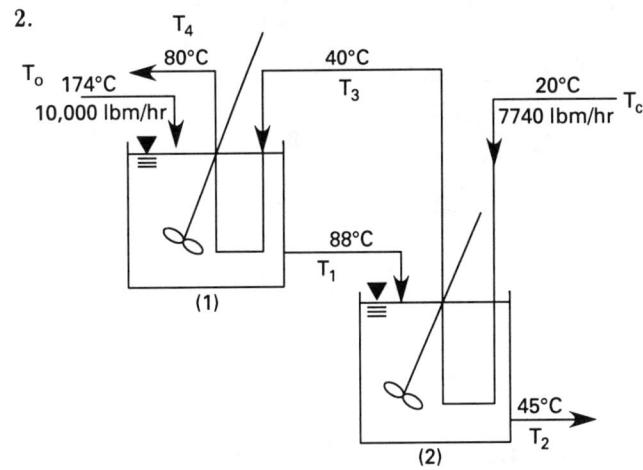

W_a = acid flow, lbm/hr
c_a = acid specific heat = 0.36 BTU/lbm-°F
T_o = acid feed temperature
T_1 = acid exit tank 1 temp, °C
T_2 = acid exit tank 2 temp, °C
V = tank capacity, lbm
t = time, hr

Assume the following:

- Boundary conditions: at $t = 0$, $T_1 = 88°$C, $T_2 = 45°$C

- Δt_m = log mean $\Delta t = \dfrac{(\Delta t_a - \Delta t_b)}{\ln\left(\dfrac{\Delta t_a}{\Delta t_b}\right)}$

- Area of coil = A ft^2

- The heat capacity is constant.

Steady state conditions:

$Q_2 = $ heat removed in tank 2
$$= W_a c_a \Delta t$$
$$= (10,000)(0.36)(88°C - 45°C)\left(\frac{1.8°F}{°C}\right)$$
$$= 278,640 \text{ BTU/hr}$$

$$Q_2 = U_2 A_2 \Delta t_m = 130 A_2 \left(\frac{48 - 25}{\ln \frac{48}{25}}\right)$$
$$= 4583.6 A_2$$

$$A_2 = \frac{278,640}{4583.6} = 60.79 \text{ ft}^2$$

$Q_1 = W_a c_a \Delta t$
$$= (10,000)(0.36)(174°C - 88°C)(1.8)$$
$$= 557,280 \text{ BTU/hr}$$

$$Q_1 = U_1 A_1 \Delta t_m = 200 A_1 \left(\frac{94 - 48}{\ln \frac{94}{48}}\right)$$
$$= 13,688.56 A_1$$

$$A_1 = \frac{557,280}{13,688.6} = 40.71 \text{ ft}^2$$

(a) The energy balance in tank 1 is

$$W_a c_a T_o - W_a c_a T_1 = \frac{d}{dt}(V_1 c_a T_1)$$

The energy balance in tank 2 is

$$W_a c_a T_1 - W_a c_a T_2 = \frac{d}{dt}(V_2 c_a T_2)$$

Use the following simultaneous first order differential equations to solve for tank 1 first.

$$T_o - T_1 = \left(\frac{V_1}{W_a}\right)\left(\frac{dT_1}{dt}\right)$$
$$T_1 - T_2 = \left(\frac{V_2}{W_a}\right)\left(\frac{dT_2}{dt}\right)$$

$$\int \frac{dT_1}{T_o - T_1} = -\ln(T_o - T_1)$$
$$= \int \frac{W_a}{V_1} dt = \frac{W_a t}{V_1} + c$$

$$T_o - T_1 = k e^{-\frac{W_a t}{V_1}}$$

Boundary conditions: $t = 0$, $T_1 = 88°C$, $k = 86$

$$T_o - T_1 = 86 e^{-\frac{W_a t}{V}}$$

Since $\frac{W_a}{V} = 1$, at $t = 1$,

$$T_1 = T_o - 86 e^{-1}$$
$$= 142.4°C \text{ after one hour cooling off}$$

Tank 1: $T_1 = T_o - 86 e^{-\frac{W_a t}{V_1}}$

Substitute into the differential equation for tank 2:

$$T_o - 86 e^{-\frac{W_a t}{V_1}} - T_2 = \left(\frac{V_2}{W_a}\right)\left(\frac{dT_2}{dt}\right)$$
$$\frac{dT_2}{dt} + \frac{W_a}{V_2} T_2 = \frac{W_a}{V_2}\left(T_o - 86 e^{-\frac{W_a t}{V_1}}\right)$$

The form is
$$\frac{dy}{dx} + p(x) y = Q(x)$$

The solution is
$$y = c e^{-\int p dx} + e^{-\int p dx} \int e^{\int p dx} Q \, dx$$
$$T_2 = c e^{-\frac{W_a t}{V_2}} + e^{-\frac{W_a t}{V_2}} \left[\int e^{\frac{W_a t}{V_2}} \left(\frac{W_a}{V_2}\right)\left(T_o - 86 e^{-\frac{W_a t}{V_1}}\right) dt\right]$$

$$\frac{W_a}{V_1} = \frac{W_a}{V_2} = 1$$

$$T_2 = c e^{-t} + e^{-t}\left[\int e^t (T_o - 86 e^{-t}) dt\right]$$
$$= c e^{-t} + T_o - 86 t e^{-t}$$

At $t = 0$, $T_2 = 45°C$.

$$c + 174°C = 45°C$$
$$c = -129°C$$
$$T_2 = T_o - (86 t + 129) e^{-t}$$

At $t = 1$,
$$T_2 = 174°C - (86 + 129) e$$
$$= \boxed{94.9°C}$$

(b) Let the new water $= W$ lbm/hr.

$T_3 = $ water exit tank 2
$T_4 = $ water exit tank 1
$c_w = $ water heat capacity
$T_c = $ water inlet temperature

The energy balance of acid and water in tank 1 is

$$Wc_wT_3 + W_ac_aT_o - (Wc_wT_4 + W_ac_aT_1) = V_1c_a\frac{dT_1}{dt}$$

The energy balance of acid and water in tank 2 is

$$Wc_wT_c + W_ac_aT_1 - (Wc_wT_3 + W_ac_aT_2) = V_2c_a\frac{dT_2}{dt}$$

The heat transfer to tank 2 is

$$Wc_w(T_3 - T_c) = U_2A_2\frac{(T_1 - T_3) - (T_2 - T_c)}{\ln\frac{(T_1 - T_3)}{(T_2 - T_c)}}$$

The heat transfer to tank 1 is

$$Wc_w(T_4 - T_3) = U_1A_1\frac{(T_o - T_4) - (T_1 - T_3)}{\ln\frac{T_o - T_4}{T_1 - T_3}}$$

To simplify, let $\alpha = e^{\frac{-U_1A_1}{Wc_w}}$ and $\beta = e^{\frac{-U_2A_2}{Wc_w}}$

$$\alpha = \frac{T_1 - T_4}{T_1 - T_3}$$

$$\beta = \frac{T_2 - T_3}{T_2 - T_c}$$

or $T_2(1 - \beta) = T_3 - \beta T_c$

$T_1(1 - \alpha) = T_4 - \alpha T_3$

Four simultaneous equations result.

$$T_3 + c_aT_o - (T_4 + c_aT_1) = c_a\frac{dT_1}{dt} \quad \text{[Eq. 1]}$$

$$T_c + c_aT_1 - (T_3 + c_aT_2) = c_a\frac{dT_2}{dt} \quad \text{[Eq. 2]}$$

$$T_1(1 - \beta) = T_3 - \beta T_c \quad \text{[Eq. 3]}$$

$$T_o(1 - \alpha) = T_4 - \alpha T_3 \quad \text{[Eq. 4]}$$

Eliminate T_4 by substituting Eq. 4 into Eqs. 2 and 1, and eliminate T_3 by substituting Eq. 3 into Eqs. 1 and 2.

$$(1 - \alpha)(1 - \beta)T_2 + (1 - \alpha)\beta T_c + c_a(T_o - T_1)$$
$$- (1 - \alpha)T_1$$
$$= c_a\frac{dT_1}{dt} \quad \text{[Eq. 5]}$$

$$T_c + c_a(T_1 - T_2) - (1 - \beta)T_2 - \beta T_c$$
$$= c_a\frac{dT_2}{dt} \quad \text{[Eq. 6]}$$

Differentiating the above equation,

$$c_a\frac{dT_1}{dt} = c_a\frac{d^2T_2}{dt^2} + (1 - \beta + c_a)\frac{dT_2}{dt} \quad \text{[Eq. 7]}$$

Using Eq. 6 to eliminate T_1 in Eq. 7 results in Eq. 8,

$$c_a^2\frac{d^2T_2}{dt^2} + (2c_a + 2 - \alpha - \beta)c_a\frac{dT_2}{dt} + [c_a^2 + c_a(1 - \alpha\beta)$$
$$+ (1 - \alpha)(1 - \beta)]T_2$$
$$= [c_a(1 - \alpha\beta) + (1 - \alpha)(1 - \beta)]T_c + c_a^2T_o$$

$$\frac{U_1A_1}{W} = \frac{(200)(40.71)}{10{,}000} = 0.8141 \text{ BTU/lbm-°F}$$

$$\alpha = 0.44299 \text{ BTU/lbm-°F}$$

$$\frac{U_2A_2}{W} = \frac{(130)(60.79)}{10{,}000} = 0.79027 \text{ BTU/lbm-°F}$$

$$\beta = 0.4537 \text{ BTU/lbm-°F}$$

Equation 8 becomes

$$\frac{d^2T_2}{dt^2} + 6.06\frac{dT_2}{dt}2 + 7.7T_2 = 308$$

This is a second order differential equation with constant coefficients.

$$T_2 = Ae^{-4.25t}Be^{-1.81t} + 40$$

From the boundary conditions, $A = 0.6$, $B = 54.3$.

$$T_2 = 0.6e^{-4.25t} + 54.3e^{-1.81t} + 40$$

At $t = 1$ hour,

$$T_2 = 0.6e^{-4.25t} + 54.3e^{-1.81t} + 40$$

$$= \boxed{48.9°\text{C}}$$

3. Use the three point method.

form: $y = a + bx + \dfrac{c}{x^2}$

$$b = \frac{fo - pes^2t^2}{fm - nes^2t^2}$$

Use the last three data points.

$m = x_2 - x_1 = 180 - 100 = 80$

$n = x_3 - x_2 = 320 - 180 = 140$

$o = y_2 - y_1 = 1.3996 - 1.3779 = 0.0217$

$p = y_3 - y_2 = 1.4391 - 13996 = 0.0395$

Since T is measured in degrees Rankine,

$$s = \frac{x_2}{x_1} = \frac{180 + 460}{100 + 460} = 1.1429$$

$$t = \frac{x_3}{x_2} = \frac{320 + 460}{180 + 460} = 1.2188$$

$f = x_3^2 - x_2^2 = (780)^2 - (640)^2 = 198{,}800$

$e = x_2^2 - x_1^2 = (640)^2 - (560)^2 = 96{,}000$

$$b = \frac{(198{,}800)(0.0217) - (0.0395)(96{,}000)\left[(1.1429)(1.2188)\right]^2}{(198{,}800)(80) - (140)(96{,}000)\left[(1.1429)(1.2188)\right]^2}$$

$$= 2.991 \times 10^{-4}$$

$$c = \frac{(bm-o)x_1^4}{es^2}$$

$$= \frac{[2.991 \times 10^{-4}(80) - 0.0217](560)^4}{(96{,}000)(1.1429)^2}$$

$$= 1747.3$$

$$a = y_1 - bx_1 - \frac{c}{x_1^2}$$

$$= 1.3779 - [(2.991 \times 10^{-4})(560)] - \frac{1747.3}{(560)^2}$$

$$= 1.2048$$

For $t = 250°\text{F}$, $T = 250 + 460 = 710°\text{R}$.

$$c_p = 1.2048 + [(2.991 \times 10^{-4})(710)] + \frac{1747.3}{(710)^2}$$

$$= \boxed{1.4206 \text{ BTU/lbm-}°\text{F}}$$

4. $kt = \dfrac{1}{c_a} - \dfrac{1}{c_{a_o}}$

Let $\dfrac{1}{c_a} = y$, $\dfrac{1}{c_{a_o}} = b$, $t = x$, $k = m$.

$$mx = y - b$$
$$y = mx + b \text{ (linear)}$$

c_a	y	x	xy	x^2
2.2502	0.4444	2	0.8888	4
1.5888	0.6294	7	4.4058	49
1.4217	0.7034	9	6.3304	81
1.1745	0.8514	13	11.0685	169
0.5298	1.8875	41	77.3877	1681
0.3753	2.6645	62	165.2012	3844
	7.1806	134	265.2825	5828

$$m = \frac{n\Sigma xy - \Sigma x \Sigma y}{n\Sigma x^2 - (\Sigma x)^2}$$

$$= \frac{(6)(265.2825) - (134)(7.1806)}{(6)(5828) - (134)^2}$$

$$m = k = 0.0370 \; \ell/\text{mol·min}$$

$$b = \bar{y} - m\bar{x}$$

$$= \frac{7.1807}{6} - \left[(0.0370)\left(\frac{134}{6}\right)\right]$$

$$= 0.3705$$

$$c_{a_o} = \frac{1}{b} = \frac{1}{0.3705}$$

$$= \boxed{2.70 \text{ mol}/\ell}$$

SOLUTIONS FOR CHAPTER 2
STOICHIOMETRY, HEAT, AND MATERIAL BALANCES

1. x = lbm clinker/100 lbm coal
   $6 = 0.9x$
   $x = 6.67$ lbm clinker/100 lbm coal
   $= 0.67$ lbm carbon/100 lbm coal
   $= 0.67\%$ unburned carbon

   percent burned $= 100 - 0.67$
   $= \boxed{99.33\% \text{ (coal basis)}}$
   $= \left(1.00 - \dfrac{0.0067}{0.79}\right)(100)$
   $= \boxed{99.15 \text{ (carbon basis)}}$

2. Since the wet basis is

   $x = \dfrac{\text{lbm H}_2\text{O evaporated}}{\text{lbm dry wood}}$

   $= \dfrac{1.00 - \dfrac{0.6}{0.8}}{0.6}$

   $= \boxed{0.4167 \text{ lbm H}_2\text{O evaporated/lbm dry wood}}$

3. $K_p = \dfrac{p_{NH_3}^2}{p_{N_2} p_{H_2}^3} = \dfrac{y_{NH_3}^2}{y_{N_2} y_{H_2}^3} P^{-2}$

   Let x represent the number of lbmoles of NH_3 formed in the reaction.

   total moles $= \left(1 - \dfrac{x}{2}\right) + \left(3 - \dfrac{3x}{2}\right) + x = 4 - x$

   $y_{NH_3} = \dfrac{x}{4-x} \quad y_{N_2} = \dfrac{1 - \dfrac{x}{2}}{4-x} \quad y_{H_2} = \dfrac{3 - \dfrac{3x}{2}}{4-x}$

   The total lbmoles of NH_3 formed is 0.148. Therefore,

   $y_{NH_3} = 0.0384 \quad y_{N_2} = 0.2404 \quad y_{H_2} = 0.7212$

   $K_o = \dfrac{(0.0384)^2}{(0.2404)(0.7212)^3} 10^{-2}$

   $= \dfrac{1.635 \times 10^{-4}}{\text{atm}^2}$

   $p_{H_2} = Py_{H_2} = (10)(0.7212)$
   $= \boxed{7.21 \text{ atm}}$

4.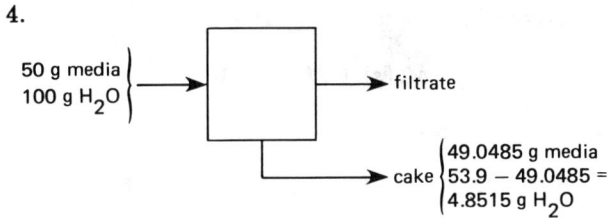

   The material balance is

	feed	cake	filtrate
H_2O	100	4.8515	95.1485
media	50	49.0485	0.9515

   solubility $= \left(\dfrac{0.9515}{95.1485}\right)(100)$

   $= \boxed{1.00002 \text{ g media/100 g H}_2\text{O}}$

5.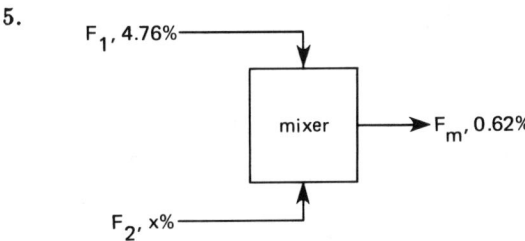

   Assume that $x = 0$.

   $F_1 + F_2 = F_m$
   $F_1(4.76) = (F_1 + F_2)(0.62)$
   $\dfrac{F_2}{F_1} = \dfrac{4.76 - 0.62}{0.62} = \boxed{6.677}$

6.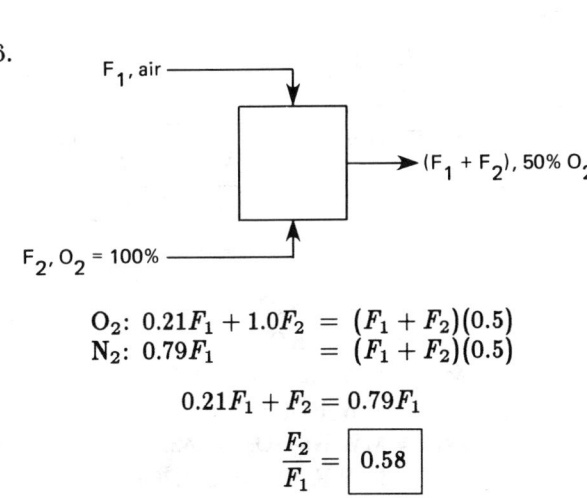

   $O_2: 0.21F_1 + 1.0F_2 = (F_1 + F_2)(0.5)$
   $N_2: 0.79F_1 = (F_1 + F_2)(0.5)$

   $0.21F_1 + F_2 = 0.79F_1$

   $\dfrac{F_2}{F_1} = \boxed{0.58}$

7. M_s = MW $Na_2SO_4 \cdot 10H_2O$ = 322
    M_c = MW $Na_2CO_3 \cdot 10H_2O$ = 286
    c = moles carbonate
    s = moles sulfate

The overall balance is
$$M_s(s) + M_c(c) = 100$$

The water balance is
$$(18)(10s) + (18)(10c) = 100 - 39.6$$
$$s + c = 0.3356$$

$$c = \left(\frac{100}{M_s} - s\right)\left(\frac{M_s}{M_c}\right)$$
$$= \left(\frac{M_s}{M_c - M_s}\right)\left(\frac{100}{M_s} - 0.3356\right)$$
$$c = 0.2239$$
$$s = 0.1116$$
$$\frac{c}{s} = \frac{0.2239}{0.1116}$$
$$= \boxed{2.006}$$

8.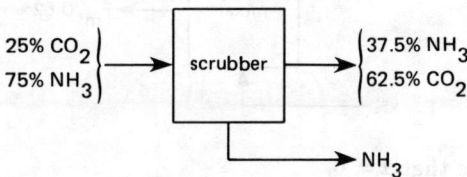

Assume:
- 100 lbmoles feed
- CO_2 tie compound

Let x represent the number of lbmoles of NH_3 in the tail gas.
$$x = (25)\left(\frac{0.375}{0.625}\right) = 15 \text{ moles}$$

$$\text{percent } NH_3 \text{ recovered} = \left(\frac{75-15}{75}\right)(100)$$
$$= \boxed{80\%}$$

9. s = moles NaCl
    p = moles KCl
    M_s = MW NaCl = 58.443
    M_p = MW KCl = 74.555
    M_1 = MW Na_2SO_4 = 142.041
    M_2 = MW K_2SO_4 = 174.266

$NaCl + KCl + H_2SO_4 \longrightarrow \frac{1}{2}Na_2SO_4 + \frac{1}{2}K_2SO_4 + HCl$
s moles p moles excess $\frac{s}{2}$ moles $\frac{p}{2}$ moles

$$s(M_s) + p(M_p) = 1$$
$$\left(\frac{s}{2}\right)(M_1) + \left(\frac{p}{2}\right)(M_2) = 1.2$$
$$s = 0.0116 \text{ moles NaCl}$$
$$p = 0.0043 \text{ moles KCl}$$

$$(0.0116)(58.443) = \boxed{0.6779 \text{ lbm NaCl}}$$
$$(0.0043)(74.555) = \boxed{0.3221 \text{ lbm KCl}}$$

10. MW feldspar = 556.67 tons/tonmole
    MW clay = 258.16 tons/tonmole

    1 tonmole feldspar $\longrightarrow$ 1 tonmole clay

    $$\frac{\text{tons clay}}{\text{tons feldspar}} = \frac{258.16}{556.67} = \boxed{0.4638}$$

11.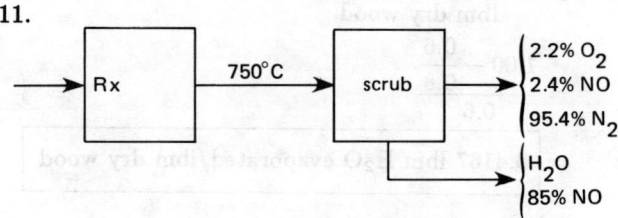

Use a basis of 100 moles of gas from the scrub.

	exit, moles	
	reactor	scrubber
O_2	2.2	2.2
NO	16.0 = $\frac{2.4}{0.15}$	2.4
N_2	95.4	95.4
NH_3	0	0

moles NH_3 reacted in desired reaction = 16
moles NH_3 lost in side reaction = x
moles NH_3 in feed = $16 + x$
moles air in feed = y

The N_2 balance is
$$0.79y + \left(\frac{1}{2}\right)(16+x) = \frac{16}{2} + 95.4$$
$$0.79y + 0.5x = 95.4$$

The O_2 balance is
$$0.21y = \frac{16}{2} + (0.75)(16+x) + 2.2$$
$$0.21y - 0.75x = 22.2$$
$$x = 3.58; \quad y = 118.5 \text{ moles air}$$

(a) The percent lost in the side reaction is

$$\left(\frac{3.58}{16+3.58}\right)(100) = \boxed{18.3\%}$$

(b) Assume that the excess air is based on the desired reaction.

$$\text{feed} = 118.5 \text{ moles air or } 24.89 \text{ moles O}_2$$

$$\text{theoretical O}_2 = (19.58)\left(\frac{5}{4}\right) = 24.48$$

$$\text{percent excess} = \left(\frac{24.89 - 24.48}{24.48}\right)(100)$$

$$= \boxed{1.70\%}$$

12.

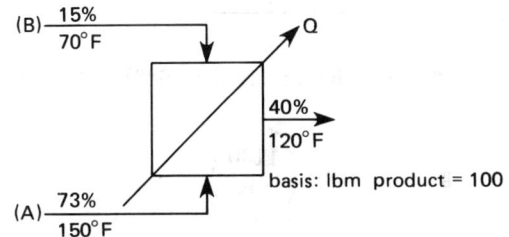

The basis is 100 lbm product.

$$\text{overall:} \quad A + B = 100$$

$$\text{NaOH:} \quad 0.73A + 0.15B = 40$$

$$A = 43.1, \quad B = 56.9$$

The heat balance is

$$340A + 30B = (107)(40) + Q$$

$$Q = (340)(43.1) + (30)(56.9) - (107)(40)$$

$$= \boxed{5661 \text{ BTU}/100 \text{ lbm product}}$$

13. The molecular weights are

$$\text{NaCN} = 49.02$$
$$\text{NaOH} = 40$$
$$\text{Cl}_2 = 70.91$$
$$\text{NaCNO} = 65.02$$

$$\frac{\text{NaCN}}{\text{day}} = \left(3600 \frac{\text{gal}}{\text{day}}\right)\left(8.33 \frac{\text{lbm}}{\text{gal}}\right)(0.0164)$$

$$= 492 \text{ lbm/day}$$

$$\frac{\text{NaCNO}}{\text{day}} = (492)\left(\frac{65.02}{49.02}\right)$$

$$= 652 \text{ lbm/day}$$

For the first step,

$$\text{lbm NaOH} = (1.15)(492)\left[\frac{(2)(40)}{49.02}\right]$$

$$= 922 \text{ lbm/day}$$

For the second step,

$$\text{lbm NaOH} = (1.15)(652)\left[\frac{(4)(40)}{(2)(65.02)}\right]$$

$$= 922 \text{ lbm/day}$$

In the first step,

$$922 = 802 + 120 \text{ excess}$$

In the second step,

$$922 = 802 + 120 \text{ excess}$$

(a) Use the same excess for both steps.

$$\text{total NaOH needed} = 802 + 802 + 120$$

$$= \boxed{1724 \text{ lbm NaOH/day}}$$

For Cl_2, the first step is

$$\text{lbm Cl}_2 = (1.20)(492)\left(\frac{70.91}{49.02}\right) = 853 \text{ lbm}$$

$$(711 \text{ lbm} + 142 \text{ lbm excess})$$

For Cl_2, the second step is

$$\text{lbm Cl}_2 = (1.20)(652)\left[\frac{(3)(70.91)}{(2)(65.02)}\right]$$

$$= 1278 \text{ lbm}$$

$$(1065 \text{ lbm} + 213 \text{ lbm excess})$$

Use 213 lbm excess for both steps.

$$\text{lbm Cl}_2 = 711 + 1065 + 213$$

$$= \boxed{1989 \text{ lbm Cl}_2/\text{day}}$$

$$\text{NaOH min weight \%} = \left[\frac{1724}{(9600)(8.33)}\right](100)$$

$$= \boxed{2.16\%}$$

(b) The heat balance is $\dot{m}c_p \Delta t = Q$.

$$(3600)(1)(90°F - 85°F) + (9600)(1)(120°F - 85°F)$$

$$= x(1)(85°F - 75°F)$$

$$x = \frac{\text{gal}}{\text{day}} \text{ cooling water}$$

$$= \boxed{35{,}400 \text{ gal/day}}$$

(c) NaCl formed

In the first step,
$$(492)\left(\frac{116.91}{49.02}\right) = 1173 \text{ lbm}$$

In the second step,
$$(652)\left[\frac{(6)(58.45)}{(2)(65.02)}\right] = 1758 \text{ lbm}$$

$$\text{total NaCl formed} = 1173 + 1758$$
$$= 2931 \text{ lbm/day}$$

$$\text{NaCl} = (10)^6 \left[\frac{2931}{(48,600)(8.33)}\right]$$
$$= \boxed{7300 \text{ ppm}}$$

14.

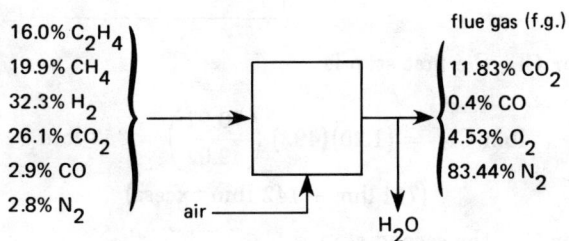

(Flue gas is abbreviated f.g.)

Choose a basis of 100 lbmoles of feed.

	elements: feed, lbmoles			
	C	O_2	H_2	N_2
C_2H_4	32		32	
CH_4	19.9		39.8	
H_2			32.3	
CO_2	26.1	26.1		
CO	2.9	1.45		
N_2				2.8
	80.9	27.55	104.1	2.8

$$\text{moles } H_2O \text{ in f.g.} = \text{moles } H_2 \text{ in feed}$$
$$= 104.1 \frac{\text{moles } H_2O \text{ in f.g.}}{100 \text{ moles feed}}$$

The amount of carbon used to determine moles of dry flue gas is

$$\text{moles dry f.g} = \frac{80.9}{(0.1183 + 0.004)}$$
$$= 661.5$$

$$\text{moles wet f.g.} = 661.5 + 104.1$$
$$= \frac{765.6 \text{ moles f.g.}}{100 \text{ moles feed}}$$

Let x represent the number of lbmoles air from the N_2 balance.

$$0.79x + 2.8 = (0.8344)(661.5)$$
$$x = 695.1 \text{ lbmoles air}$$

(a) mole ratio = volume ratio

$$\text{air} = \boxed{6.951 \text{ ft}^3 \text{ air/ft}^3 \text{ feed}}$$

(b) At 670°F (1130°R) and feed at 68°F (528°R),

$$\text{f.g.} = (7.656)\left(\frac{1130}{528}\right)$$
$$= \boxed{16.385 \text{ ft}^3 \text{ f.g./ft}^3 \text{ at } 68°F}$$

15. $M_s = \text{MW Na}_2SO_4 = 142.04$
$M_w = \text{MW H}_2O = 18.01$
$M_D = \text{MW Na}_2SO_4 \cdot 10 H_2O = 322.14$
$x_s = \text{moles NaSO}_4 \text{ in solid}$

$$\text{solubility} = \frac{\left(\frac{10}{M_s}\right) - x_s}{\left(\frac{90}{M_w}\right) - 10x_s} = 0.00634$$

To determine $x_s M_D$,

$$x_s M_D = M_D \left(\frac{\frac{10}{M_s} - \left(\frac{90}{M_w}\right)(0.00634)}{0.9366}\right)$$

$$= (322.14) \left(\frac{\frac{10}{142.04} - \left(\frac{90}{18.01}\right)(0.00634)}{0.9366}\right)$$

$$= \boxed{13.318 \text{ lbm solid}/100 \text{ lbm mix}}$$

STOICHIOMETRY, HEAT, AND MATERIAL BALANCES

16.

	feed, lbmoles	100 lbmole basis
	C	H_2
C_7H_{16}	350	400
C_8H_{18}	400	450
	750	850

$$O_2 \text{ required} = (1.1)\left[750 + \left(\frac{1}{2}\right)(850)\right]$$
$$= 1292.5 \text{ lbmole } O_2/100 \text{ lbmole fuel}$$
$$\text{air} = \frac{1292.5}{0.21}$$
$$= 6154.8 \text{ lbmole air}/100 \text{ lbmole fuel}$$

100 lbmoles fuel $= (50)(100 + 114)$
$$= 10{,}700 \text{ lbm fuel}$$
$$\text{air} = (6154.8)\left(\frac{100}{10{,}700}\right)$$
$$= 57.52 \text{ lbmoles air}/100 \text{ lbm fuel}$$

(a) $\dfrac{\text{ft}^3 \text{ air}}{100 \text{ lbm fuel}} = (57.52)\left(359 \dfrac{\text{ft}^3}{\text{lbmole}}\right)$
$$\times \left(\frac{530°R}{492°R}\right)\left(\frac{14.7 \text{ psi}}{14.9 \text{ psi}}\right)$$
$$= 21{,}946 \text{ ft}^3 \text{ air}/100 \text{ lbm fuel}$$

Choose a basis of 100 lbmoles of fuel.

CO_2: 750
H_2O: 850
O_2: $1292.5 - 750 - \left(\frac{1}{2}\right)(850) = 117.5$
N_2: $(6154.8)(0.79) = 4862.3$
total $= 6579.8$

(b) $CO_2 = \dfrac{750}{6579.8} = 11.4\%$

$H_2O = \dfrac{850}{6579.8} = 12.9\%$

$O_2 = \dfrac{117.5}{6579.8} = 1.79\%$

$N_2 = 73.91\%$

17.

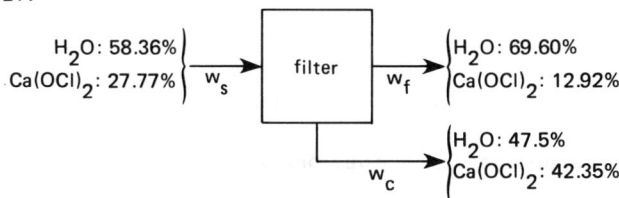

Assume:
- The remaining percent is impurities.
- The precipitate is pure $Ca(OCl)_2 \cdot 2H_2O$.
- The basis is 100 lbm of feed.

The H_2O balance is
$$58.36 = 0.696w_f + 0.475w_c$$
$$w_f + w_c = 100$$
$$w_f = 49.14 \text{ lbm}$$
$$w_c = 50.86 \text{ lbm}$$

(a) The slurry consists of pure $Ca(OCl)_2$ and its mother liquor.
$$w_{\text{mother liquor}} + w_{\text{solid}} = 100$$

The H_2O balance is
$$0.7128 w_{\text{mother liquor}} + \frac{36.04}{179.02} w_{\text{solid}} = 58.36$$
$$w_{\text{mother liquor}} = 74.74 \text{ lbm}$$
$$w_{\text{solid}} = 25.26 \text{ lbm}$$
$$\text{slurry} = \boxed{25.26\% \text{ solid}}$$

(b) $\dfrac{\text{lbm solid lost}}{100 \text{ lbm filtrate}} =$
$$\left[12.92 - \left(\frac{10.2 \text{ lbm } Ca(OCl)_2 \text{ dissolved}}{71.28 \text{ lbm } H_2O}\right)\right.$$
$$\left. \times (69.6 \text{ lbm } H_2O)\right]\left(\frac{179.02}{142.98}\right)$$
$$= 3.71 \frac{\text{lbm solid lost}}{100 \text{ lbm filtrate}}$$
$$= \boxed{3.71\% \text{ solid in filtrate}}$$

(c) $\left(\dfrac{3.71 \text{ lbm solid}}{100 \text{ lbm filtrate}}\right)\left(\dfrac{49.14 \text{ lbm filtrate}}{100 \text{ lbm slurry}}\right)\left(\dfrac{100 \text{ lbm slurry}}{25.26 \text{ lbm solid}}\right)$
$$= \frac{7.22 \text{ lbm solid lost}}{100 \text{ lbm solid in slurry}}$$
$$\text{percent lost} = \boxed{7.22\%}$$

18. $C_3H_8 + (5+15)O_2 + (5+15)\left(\dfrac{0.79}{0.21}\right)N_2 \longrightarrow$
$$3CO_2 + 4H_2O + 15O_2 + (20)\left(\frac{0.79}{0.21}\right)N_2$$

The basis is
$$1 \text{ mole fuel} = 44.092 \text{ lbm}$$

The products of combustion are

3 lbmoles CO_2: 132 lbm
15 lbmoles O_2: 480 lbm
$(20)\left(\dfrac{0.79}{0.21}\right)$ lbmoles N_2: 2107 lbm
4 lbmoles H_2O: 72 lbm

$$LHV = 19{,}944 \text{ BTU/lbm}$$

Since hot air is being used, the added sensible heat must be added to LHV to find TFT.

Assume that TFT = 2000°F.

	m	c_p	mc_p
CO_2	132	0.275	36.3
H_2O	72	0.515	37.08
O_2	480	0.25	120
N_2	2107	0.27	568.9
			762.28

$$\text{lbm air used (300\% excess)} = 2107 + (20)(32)$$
$$= 2747 \text{ lbm air}$$

$$c_p \text{ air} = 0.241$$
$$\Delta t = 260 - 77 = 183°F$$

$$\text{sensible heat from air} = (2747)(0.241)(183)$$
$$= 121{,}151 \text{ BTU}$$

$$TFT = \frac{(44.092)(19{,}944) + 121{,}151}{762.28} + 77$$
$$= 1390°F$$

Assume TFT = 1400°F.

	m	c_p	mc_p
CO_2	132	0.26	34.3
H_2O	72	0.5	36
O_2	480	0.24	115.2
N_2	2107	0.26	547.8

$$\Sigma mc_p = 733.3$$

$$TFT = \frac{(44.092)(19{,}944) + 121{,}151}{733.3} + 77$$
$$= 1441°F$$
$$= \frac{1390 + 1441}{2}$$
$$= \boxed{1416°F}$$

19. h = height in inches
c = cost in dollars
d = diameter in inches

$$c = \frac{(2\pi d^2)(7.00)}{(4)(144)} + \frac{25t\pi dh}{144}$$
$$= 0.07635d^2 + 0.5454 dht$$

$$t = \frac{pd}{24{,}000}$$
$$p = \rho h = \frac{(2)(62.4)}{1728} h$$
$$= 0.07222h$$
$$t = 3.009 \times 10^{-6} dh$$
$$h = \frac{(20{,}000)(231)}{\frac{\pi d^2}{4}}$$
$$= \frac{5.882 \times 10^6}{d^2}$$
$$c = 0.07635d^2 + \frac{5.678 \times 10^7}{d^2}$$

Find the minimum, differentiate, then set to 0.

$$\left(\frac{d}{dd}\right)(c) = 0$$
$$= 0.1527d - \frac{1.1356 \times 10^8}{d^3}$$
$$d = 165 \text{ in}$$
$$h = 216.1 \text{ in}$$
$$t = 0.107 \text{ too thin } (t = 0.25)$$

With $t = 0.25$,

$$c = 0.07635d^2 + \frac{8.02 \times 10^5}{d}$$
$$\left(\frac{d}{dd}\right)(c) = 0$$
$$= 0.1527d - \frac{8.02 \times 10^5}{d^2}$$
$$d = 173.8 \text{ in (or 14.5 ft)}$$
$$h = \frac{5.882 \times 10^6}{d^2}$$
$$= \boxed{194.7 \text{ in (or 16.2 ft)}}$$

20.
$$CaCO_3 \longrightarrow CO_2 + CaO \quad \text{[Eq. 1]}$$
$$CH_4 + 2O_2 \longrightarrow CO_2 + 2H_2O \quad \text{[Eq. 2]}$$
$$CH_4 + \tfrac{3}{2}O_2 \longrightarrow CO + 2H_2O \quad \text{[Eq. 3]}$$

Assume that the missing percentage of gas is O_2.

	moles	moles O_2	moles C
CO_2	20.4	20.4	20.4
CO	0.4	0.2	0.4
O_2	2.1	2.1	
N_2	77.1		
	100.0	22.7	20.8

$$O_2 \text{ from air} = (77.1)\left(\frac{0.79}{0.21}\right)$$
$$= 20.5 \frac{\text{lbmoles } O_2}{100 \text{ lbmoles f.g.}}$$

From Eq. 3,

$$CH_4 \text{ converted to } CO = 0.4 \frac{\text{lbmoles } CH_4}{100 \text{ lbmoles f.g.}}$$

$$O_2 \text{ consumed in Eq. 3} = 0.6 \frac{\text{lbmoles } O_2}{100 \text{ lbmoles f.g.}}$$

From Eq. 2,

$$O_2 \text{ available} = 20.5 - 2.1 - 0.6$$
$$= 17.8 \frac{\text{lbmoles } O_2}{100 \text{ lbmoles f.g.}}$$

$$CH_4 \text{ burned to } CO_2 = \frac{17.8}{2}$$
$$= 8.9 \frac{\text{lbmoles } CH_4}{100 \text{ lbmoles f.g.}}$$

$$\text{total } CH_4 \text{ burned} = 8.9 + 0.4$$
$$= 9.3 \frac{\text{lbmoles } CH_4}{100 \text{ lbmoles f.g.}}$$

$$CO_2 \text{ produced from calcine} = 20.4 - 8.9$$
$$= 11.5 \frac{\text{lbmoles } CO_2}{100 \text{ lbmoles f.g.}}$$

From Eq. 1,

$$CaO \text{ produced} = 11.5 \frac{\text{lbmoles } CaO}{100 \text{ lbmoles f.g.}}$$

$$\text{actual } CH_4 \text{ burned} = \left(\frac{29{,}000 \text{ ft}^3}{359 \frac{\text{ft}^3}{\text{lbmole}}}\right)\left(\frac{492°R}{520°R}\right)$$
$$= 76.43 \text{ lbmole/hr}$$

$$\text{moles f.g. produced} = (76.43)\left(\frac{100}{9.3}\right)$$
$$= 821.8 \text{ lbmole f.g./hr}$$

$$CaO \text{ produced} = \left(\frac{11.5}{100}\right)(821.8)(56)$$
$$= \boxed{5293 \text{ lbm } CaO/hr}$$

21.

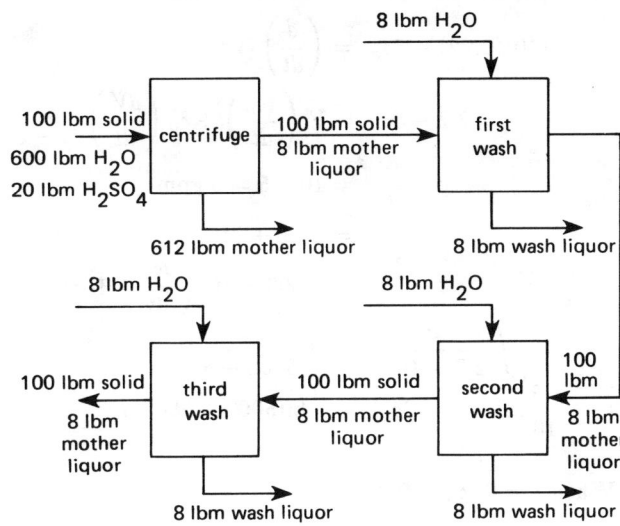

(a) In the unwashed cake,

$$\text{mother liquor acid} = \left(\frac{20}{620}\right)(8)$$
$$= 0.2581 \text{ lbm } H_2SO_4$$

For the first wash, the percent acid in the cake (dry basis) is

$$\left[\frac{(0.2)(0.2581)}{100 + (0.2)(0.2581)}\right](100) = \boxed{0.0516\%}$$

(b) $\quad \text{acid in second wash} = (0.2)(0.2)(0.2581)$
$$= 0.010324 \text{ lbm}$$

The percent of acid in the second wash cake is

$$\left[\frac{0.010324}{(100 + 0.010324)}\right](100) = 0.01032\%$$

$$\boxed{\text{Two stages are needed.}}$$

(c) acidity = 0.1032%

The total acid in the wash liquor is

$$0.2581 - 0.010324 = 0.24777 \text{ lbm } H_2SO_4$$

The total combined wash liquor is 16 lbm.

$$\text{percent acid} = \left(\frac{0.24777}{16}\right)(100)$$

$$\boxed{1.549\% \ H_2SO_4}$$

22. input − output = accumulation

$$(10)(0.22) - 5c_o = \left(\frac{d}{dt}\right)(Vc_o)$$

$$2.2 - 5c_o = V\left(\frac{dc_o}{dt}\right) + c_o\left(\frac{dV}{dt}\right)$$

$$\frac{dV}{dt} = 10 - 5 = 5 \text{ gpm}$$

$$V = 5000 + 5t$$

$$2.2 - 5c_o = (5000 + 5t)\left(\frac{dc_o}{dt}\right) + 5c_o$$

$$\int \frac{dc_o}{2.2 - 10c_o} = \int \frac{dt}{5000 + 5t}$$

$$\frac{-1}{10}\ln(2.2 - 10c_o) = \frac{1}{5}\ln(5000 + 5t) + I$$

When $t = 0$, $c_o = 0.05$.

$$I = -1.756$$

Rearrange, multiply by 10, and raise e^x.

$$10c_o = 2.2 - \left[\frac{4.228 \times 10^7}{(5000 + 5t)^2}\right]$$

(a) When $t = (5)(60) = 300$ min,

$$10c_o = 2.2 - \left[\frac{4.228 \times 10^7}{(5000 + 5(300))^2}\right]$$

$$c_o = \boxed{0.1199 \ (11.99\%)}$$

(b) If the overflow occurs at 10,000 gallons, after 12 hours,

$$\text{total volume} = 5000 + (5)[(12)(60)]$$
$$= 8600 \text{ gal, no overflow}$$

$$10c_o = 2.2 - \frac{4.228 \times 10^7}{(8600)^2}$$

$$c_o = \boxed{0.1628 \ (16.28\%)}$$

23. Since no other term except c_1 is affected by the recycle stream,

$$c_2 = c_1^*(1-q)$$
$$= \left(\frac{c_1 + c_2 R}{1 + R}\right)(1-q)$$

Rearrange to make explicit in c_2.

$$c_2 = \frac{c_1(1-q)}{1 + R - R(1-q)}$$

For $R = 0$, $q = \frac{2}{3}$.

$$c_2 = \frac{c_1\left(1 - \frac{2}{3}\right)}{(1 + 0 + 0)} = \frac{1}{3}c_1$$

For $R = 3$, $q = \frac{2}{3}$.

$$c_2 = \frac{c_1\left(1 - \frac{2}{3}\right)}{1 + 3 - (3)\left(1 - \frac{2}{3}\right)}$$

$$= \frac{1}{9}c_1$$

The removal efficiency before the recycle is

$$(100)\left(1 - \frac{1}{3}\right) = 66.7\%$$

The removal efficiency after the recycle starts is

$$(100)\left(1 - \frac{1}{9}\right) = 88.9\%$$

$$\text{percent increase} = \left(\frac{88.9 - 66.7}{66.7}\right)(100)$$

$$= \boxed{33\% \text{ increase in removal efficiency}}$$

At 95% removal,

$$c_2 = \left(\frac{1}{20}\right)c_1$$

$$\frac{c_1(1-q)}{1 + R - R(1-q)} = \left(\frac{1}{20}\right)c_1$$

$$q = \frac{2}{3}$$

$$\frac{\frac{1}{3}}{\left(1 + R - \frac{R}{3}\right)} = \frac{1}{3 + 2R} = \frac{1}{20}$$

$$R = \frac{17}{2} = 8.5$$

$$\boxed{\text{A recycle ratio of 8.5 will remove 95\% of the rubidium with a removal fraction of 2/3.}}$$

SOLUTIONS FOR CHAPTER 3
ENGINEERING ECONOMICS

1. $F = P(F/P, 6\%, 10)$
   $= (1000)(1.7908) = \boxed{\$1790.80}$

2. $P = F(P/F, 6\%, 4)$
   $= (2000)(0.7921) = \boxed{\$1584.20}$

3. $P = F(P/F, 6\%, 20)$
   $= (2000)(0.3118) = \boxed{\$623.60}$

4. $A = P(A/P, 6\%, 7)$
   $= (500)(0.1791) = \boxed{\$89.55/\text{year}}$

5. $F = A(F/A, 6\%, 10)$
   $= (50)(13.1808) = \boxed{\$659.04}$

6. If each year is independent,
   $P = F(P/F, 6\%, 1)$
   $= (200)(0.9434) = \boxed{\$188.68}$

7. $A = F(A/F, 6\%, 5)$
   $= (2000)(0.1774) = \boxed{\$354.80/\text{year}}$

8. $F = P[(F/P, 6\%, 10) + (F/P, 6\%, 8) + (F/P, 6\%, 6)]$
   $= (100)(1.7908 + 1.5938 + 1.4185)$
   $= \boxed{\$480.31}$

9. $\phi = r/k \quad n = (5)(12) = 60$
   $\phi = \dfrac{0.06}{12} = 0.005 \text{ or } 0.5\%$
   $F = P(F/P, 0.5\%, 60)$
   $= (500)(1.3489)$
   $= \boxed{\$674.45}$

10. $P = F(P/F, i\%, 7)$
    $80 = (120)(P/F, i\%, 7)$
    $(P/F, i\%, 7) = 0.6666$
    $i \approx \boxed{6\% \text{ (from the tables)}}$

11. $\text{EUAC} = (17{,}000 + 5000)(A/P, 6\%, 5) + 200$
    $\qquad - (14{,}000 + 2500)(A/F, 6\%, 5)$
    $= (22{,}000)(0.2374) + 200$
    $\qquad - (16{,}500)(0.1774)$
    $= \boxed{\$2495.70/\text{year}}$

12. Assume that the bridge will be there forever.

    Repair: consider the salvage value as a lost benefit (or cost).

    $\text{EUAC}_1 = (9000 + 13{,}000)(A/P, 8\%, 20) + 500$
    $\qquad - (10{,}000)(A/F, 8\%, 20)$
    $= (22{,}000)(0.1019) + 500 - (10{,}000)(0.0219)$
    $= \$2522.80/\text{year}$

    Replace:

    $\text{EUAC}_2 = (40{,}000)(A/P, 8\%, 25) + 100$
    $\qquad - (15{,}000)(A/F, 8\%, 25)$
    $= (40{,}000)(0.0937) + 100 - (15{,}000)(0.0137)$
    $= \$3642.50/\text{year}$

    $\boxed{\text{Repair the bridge.}}$

13. $D = \dfrac{150{,}000}{15} = \$10{,}000/\text{year}$
    $P = -I + R(1-t) - c(1-t) + D(t)$

    $P = 0$
    $= -150{,}000 + (32{,}000)(1 - 0.48)(P/A, i\%, 15)$
    $\qquad - (7530)(1 - 0.48)(P/A, i\%, 15)$
    $\qquad + (10{,}000)(0.48)(P/A, i\%, 15)$
    $150{,}000 = (16{,}640 - 3915.6 + 4800)(P/A, i\%, 15)$

    $(P/A, i\%, 15) = 8.5595$

    $i = \boxed{8\% \text{ (from the tables)}}$

14. (a) $\dfrac{1,500,000 - 300,000}{1,000,000} = \boxed{1.2}$

(b) $1,500,000 - 300,000 - 1,000,000 = \boxed{\$200,000}$

15. Assume that renovation occurs at $t = 0$.

$$\begin{aligned}
0 &= P \\
&= -(14,000 + 1000) + [(75)(12) - 150 - 250] \\
&\quad (P/A,\ 10\%,\ 10) + S(P/F,\ 10\%,\ 10) \\
&= -15,000 - (500)(6.1446) + S(0.3855)
\end{aligned}$$

$11{,}927.7 = (0.3855)S$

$S = \boxed{\$30{,}940.86}$

16. $P = A(P/A,\ i\%,\ 30)$

$(P/A,\ i\%,\ 30) = \dfrac{2000}{89.30} = 22.396$

$i = 2\%$ per month

$= (1 + \phi)^k - 1$

$= (1 + 0.02)^{12} - 1$

$= \boxed{0.2682\ (26.82\%)}$

17. (a) $SL:\ D = \dfrac{I - S}{t}$

$= \dfrac{500{,}000 - 100{,}000}{25}$

$= \boxed{\$16{,}000}$

(b) SOYD: $T = \left(\dfrac{1}{2}\right)(25)(26) = 325$

$D_1 = \left(\dfrac{25}{325}\right)(500{,}000 - 100{,}000)$

$= \$30{,}769$

$D_2 = \left(\dfrac{24}{325}\right)(400{,}000) = \$29{,}538$

$D_3 = \left(\dfrac{23}{325}\right)(400{,}000) = \boxed{\$28{,}308}$

(c) DDB: $D_1 = \left(\dfrac{2}{25}\right)(500{,}000) = \$40{,}000$

$D_2 = \left(\dfrac{2}{25}\right)(500{,}000 - 40{,}000)$

$= \$36{,}800$

$D_3 = \left(\dfrac{2}{25}\right)(500{,}000 - 40{,}000 - 36{,}800)$

$= \boxed{\$33{,}856}$

18. $\begin{aligned} P &= -12{,}000 - (1000)(P/A,\ 10\%,\ 10) \\ &\quad - (200)(P/G,\ 10\%,\ 10) \\ &\quad + (2000)(P/F,\ 10\%,\ 10) \\ &= -12{,}000 - (1000)(6.1446) - (200)(22.8913) \\ &\quad + (2000)(0.3855) \\ &= \boxed{-\$21{,}951.86} \end{aligned}$

$\begin{aligned} EUAC &= (21{,}951.86)(A/P,\ 10\%,\ 10) \\ &= (21{,}951.86)(0.1627) \\ &= \boxed{\$3571.56/\text{year}} \end{aligned}$

19. If the probability of failure for a life of N is $1/N$ each year,

$\begin{aligned} EUAC_9 &= (1500)(A/P,\ 6\%,\ 20) + \left(\dfrac{1}{9}\right)(0.35)(1500) \\ &\quad + (0.04)(1500) \\ &= (1500)\left[0.0872 + \left(\dfrac{1}{9}\right)(0.35) + 0.04\right] \\ &= \$249.13 \end{aligned}$

$\begin{aligned} EUAC_{14} &= (1600)\left[0.0872 + \left(\dfrac{1}{14}\right)(0.35) + 0.04\right] \\ &= \$243.52 \end{aligned}$

$EUAC_{30} = (1750)\left[0.1272 + \left(\dfrac{1}{30}\right)(0.35)\right] = \$243.01$

$EUAC_{52} = (1900)\left[0.1272 + \left(\dfrac{1}{52}\right)(0.35)\right] = \$254.47$

$EUAC_{86} = (2100)\left[0.1272 + \left(\dfrac{1}{86}\right)(0.35)\right] = \$275.64$

Choose the 30 year pipe since its EUAC is lowest.

20. $\text{EUAC}_7 = (0.15)(25{,}000) = \$3750.00$

$\text{EUAC}_8 = (15{,}000)(A/P, 10\%, 20) + (0.10)(25{,}000)$
$\phantom{\text{EUAC}_8} = \$4262.50$

$\text{EUAC}_9 = (20{,}000)(0.1175) + (0.07)(25{,}000)$
$\phantom{\text{EUAC}_9} = \$4100.00$

$\text{EUAC}_{10} = (30{,}000)(0.1175) + (0.03)(25{,}000)$
$\phantom{\text{EUAC}_{10}} = \$4275.00$

$\boxed{\text{It is cheapest to do nothing.}}$

21. (a) $\text{EUAC}_1 = (10{,}000)(A/P, 20\%, 1) + 2000$
$\phantom{\text{EUAC}_1 =} - (8000)(A/F, 20\%, 1)$
$\phantom{\text{EUAC}_1} = (10{,}000)(1.20) + 2000 - (8000)(1.0)$
$\phantom{\text{EUAC}_1} = \$6000.00$

$\text{EUAC}_2 = (10{,}000)(A/P, 20\%, 2) + 2000$
$\phantom{\text{EUAC}_2 =} + (1000)(A/G, 20\%, 2)$
$\phantom{\text{EUAC}_2 =} - (7000)(A/F, 20\%, 2)$
$\phantom{\text{EUAC}_2} = (10{,}000)(0.6545) + 2000 + (1000)(0.4545)$
$\phantom{\text{EUAC}_2 =} - (7000)(0.4545) = \$5818.00$

$\text{EUAC}_3 = (10{,}000)(A/P, 20\%, 3) + 2000$
$\phantom{\text{EUAC}_3 =} + (1000)(A/G, 20\%, 3)$
$\phantom{\text{EUAC}_3 =} - (6000)(A/F, 20\%, 3)$
$\phantom{\text{EUAC}_3} = (10{,}000)(0.4747) + 2000 + (1000)(0.8791)$
$\phantom{\text{EUAC}_3 =} - (6000)(0.2747) = \$5977.90$

$\text{EUAC}_4 = (10{,}000)(A/P, 20\%, 4) + 2000$
$\phantom{\text{EUAC}_4 =} + (1000)(A/G, 20\%, 4)$
$\phantom{\text{EUAC}_4 =} - (5000)(A/F, 20\%, 4)$
$\phantom{\text{EUAC}_4} = (10{,}000)(0.3863) + 2000 + (1000)(1.2742)$
$\phantom{\text{EUAC}_4 =} - (5000)(0.1863) = \$6205.70$

$\text{EUAC}_5 = (10{,}000)(A/P, 20\%, 5) + 2000$
$\phantom{\text{EUAC}_5 =} + (1000)(A/G, 20\%, 5)$
$\phantom{\text{EUAC}_5 =} - (4000)(A/F, 20\%, 5)$
$\phantom{\text{EUAC}_5} = (10{,}000)(0.3344) + 2000 + (1000)(1.6405)$
$\phantom{\text{EUAC}_5 =} - (4000)(0.1344) = \$6446.40$

$\boxed{\text{Sell at the end of the second year.}}$

(b) $\text{EUAC} = 6000 + i5000 + (5000 - 4000)$
$\phantom{\text{EUAC}} = 6000 + (0.2)(5000) + (5000 - 4000)$
$\phantom{\text{EUAC}} = \boxed{\$8000.00}$

22. The man should charge his company for only the additional business travel expenses.

maintenance: $\$200 - \$150 = \$50/\text{year}$

insurance: $\$300 - \$200 = \$100/\text{year}$

salvage: $(\$1000 - \$500)(A/F, 10\%, 5)$
$\phantom{\text{salvage:}} = \$81.90/\text{year}$

gasoline: $(5000)\left(\dfrac{0.6}{15}\right) = \$200.00/\text{year}$

$\text{cost per mile} = \dfrac{100 + 50 + 81.9 + 200}{5000} = \$0.0864$

(a) $\boxed{\text{Yes, \$0.10 per mile is adequate.}}$

(b) $(0.1)(x) = (5000)(A/P, 10\%, 5) + 250$
$ + 200 - (800)(A/F, 10\%, 5)$
$ + \left(\dfrac{x}{15}\right)(0.60)$

$0.1x = 1637.96 + 0.04x$

$x = \boxed{27{,}299 \text{ miles/year}}$

SOLUTIONS FOR CHAPTER 4
THERMODYNAMICS

1. Since no external force moves,

$$\Delta W = 0 \quad \text{and} \quad \Delta Q = 0$$

From the first law of thermodynamics,

$$\Delta U_1 + \Delta U_2 = 0$$

Since $\Delta U = nC_v \Delta T$,

$$n_1 C_v (T - T_1) + n_2 C_v (T - T_2) = 0$$

or

$$T = \frac{n_1 T_1 + n_2 T_2}{n_1 + n_2} \quad \text{[Eq. 1]}$$

For an ideal gas,

$$n_1 = \frac{p_1 V_1}{RT_1} \quad n_2 = \frac{p_2 V_2}{RT_2}$$

Substituting into Eq. 1 results in

$$T = T_1 T_2 \frac{p_1 V_1 + p_2 V_2}{(p_1 V_1 T_2 + p_2 V_2 T_1)}$$

$T_1 = 473\text{K} \quad T_2 = 573\text{K}$
$p_1 = 8 \text{ atm} \quad p_2 = 6 \text{ atm}$
$V_1 = 9 \text{ ft}^3 \quad V_2 = 1 \text{ ft}^3$

$$T = (473)(573)\left[\frac{(8)(9) + (6)(1)}{(8)(9)(573) + (6)(1)(473)}\right]$$

$$= \boxed{479.4\text{K}}$$

$$p = \frac{(n_1 + n_2)RT}{V_1 + V_2} = \frac{p_1 V_1 + p_2 V_2}{V_1 + V_2}$$

$$= \frac{(8)(9) + (6)(1)}{9 + 1} = \boxed{7.8 \text{ atm}}$$

2.
$$\Delta U = nC_v \Delta T = nC_v(T_2 - T_1)$$
$$= nC_v \left(\frac{p_2 V_2}{Rn} - \frac{p_1 V_1}{Rn}\right)$$
$$= C_v \left(\frac{p_2 V_2 - p_1 V_1}{R}\right)$$

For an ideal gas, $C_p - C_v = R$. Therefore, $C_v \sim 5$.

$$\Delta U = (5)\left[\frac{(119.7)(0.01) - (114.7)(0.01)}{10.73}\right]$$

$$= \boxed{0.0233 \text{ BTU}}$$

3. For an ideal gas,

$$C_p - C_v = R$$

$\boxed{C_p \text{ is larger.}}$

4.
$$\Delta s = \frac{\Delta Q}{T} \quad \text{(phase change)}$$
$$= \frac{96}{368.5}$$
$$= \boxed{0.2605 \text{ cal/gmol·K}}$$

$\Delta W = 0 \quad \Delta T = 0$

From the first law,

$$\Delta U = \boxed{0}$$

5. For 1 gmol,

$$\Delta H = (1)C_{p_2}\Delta T_2 - (1)\Lambda - (1)C_p\Delta T_1$$
$$= (1)(9)(-2) - (1)(1436) - (1)(18)(-2)$$
$$= \boxed{-1418 \text{ cal}}$$

$$\Delta S = \frac{\Lambda}{T} + C_{p_2}\ln\left(\frac{T_1}{T_2}\right) + C_{p_1}\ln\left(\frac{T_2}{T_1}\right)$$
$$= \frac{-1436}{273} + 0.5\ln\left(\frac{273}{271}\right) + 1.0\ln\left(\frac{271}{273}\right)$$
$$= \boxed{-5.26 \text{ cal/K}}$$

6.
$$-\frac{\Delta H}{R} = \frac{\ln K_2 - \ln K_1}{\frac{1}{T_2} - \frac{1}{T_1}}$$
$$= \frac{\ln(3992) - \ln(0.0002)}{\frac{1}{1273} - \frac{1}{873}}$$
$$= -46\,701\text{K}$$

$$\Delta H = (46\,701)(1.987) = 92\,796 \text{ cal/gmol·K}$$

$$-46\,701 = \frac{\ln(3992) - \ln K_{900}}{\frac{1}{1273} - \frac{1}{1173}}$$

$$K_{900} = \boxed{175}$$

$$I_2 \longrightarrow 2I$$

Let x represent the number of moles I_2 dissociated; 1 mole I_2 start.

$$K = \frac{(2x)^2}{1-x}$$

$$x^2 + \frac{K}{4}x - \frac{K}{4} = 0$$

$$= 4x^2 + Kx - K$$

$$x = \frac{-175 + \sqrt{(175)^2 + (16)(175)}}{8}$$

$$= \boxed{0.978}$$

7. Van der Waals' equation is

$$\left(p + \frac{a}{V^2}\right)(V - b) = RT$$

Solve for $\frac{pV}{RT}$.

$$\frac{pV}{RT} = Z = \frac{V}{V-b} - \frac{a}{RTV}$$

$$Z = \frac{1}{1 - \frac{b}{V}} - \frac{\frac{a}{RT}}{V}$$

From the binomial series of algebra,

$$\frac{1}{1-x} = 1 + x + x^2 + x^3 + \cdots \quad (x^2 < 1)$$

$$Z = 1 + \frac{b}{V} - \frac{\frac{a}{RT}}{V} + \frac{b^2}{V^2} + \frac{b^3}{V^3} + \cdots$$

The virial form is

$$\boxed{Z = 1 + \frac{b - \frac{a}{RT}}{V} + \frac{b^2}{V^2} + \frac{b^3}{V^3} + \frac{b^4}{V^4} + \cdots}$$

8. From App. C,

$$\boxed{\text{For all } T_R, \text{ if } P_R > 7.5, Z > 1.}$$

9. If $y = f(x, z)$, and $\partial y/\partial x = 0$, then $y = f(z)$ only.

If $H = f(T, p)$ and $U = f(T, V)$,

and $$dU = \left(\frac{\partial U}{\partial T}\right)_V dT + \left(\frac{\partial U}{\partial V}\right)_T dV$$

$$dH = \left(\frac{\partial H}{\partial T}\right)_p dT + \left(\frac{\partial H}{\partial p}\right)_T dp$$

It can be shown from the definition of U that

$$dU = TdS - pdV$$

From the Maxwell equations,

$$\left(\frac{\partial U}{\partial V}\right)_T = T\left(\frac{\partial p}{\partial T}\right)_V - p$$

For an ideal gas, $pV = RT$.

$$\left(\frac{\partial p}{\partial T}\right)_V = \frac{p}{T}$$

$$\left(\frac{\partial U}{\partial V}\right)_T = T\left(\frac{p}{T}\right) - p = 0 \text{ (ideal gas)}$$

$$dU = \left(\frac{\partial U}{\partial T}\right)_V dT$$

$$U = \boxed{f(T) \text{ only (ideal gas)}}$$

Likewise,

$$\left(\frac{\partial H}{\partial p}\right)_T = V - T\left(\frac{\partial V}{\partial T}\right)_p \text{ (any gas)}$$

$$\left(\frac{\partial V}{\partial T}\right)_p = \frac{V}{T} \text{ (ideal gas)}$$

$$\left(\frac{\partial H}{\partial p}\right)_T = V - T\left(\frac{V}{T}\right) = 0$$

$$H = \boxed{f(T) \text{ only (ideal gas)}}$$

By definition,

$$C_v = \left(\frac{\partial U}{\partial T}\right)_V \quad C_p = \left(\frac{\partial H}{\partial T}\right)_p$$

It can be shown that for any gas,

$$\left(\frac{\partial C_v}{\partial V}\right)_T = T\left(\frac{\partial^2 p}{\partial T^2}\right)_V$$

$$\left(\frac{\partial C_p}{\partial p}\right)_T = -T\left(\frac{\partial^2 V}{\partial T^2}\right)_p$$

For an ideal gas,

$$\left(\frac{\partial^2 p}{\partial T^2}\right)_V = 0$$

$$\left(\frac{\partial^2 V}{\partial T^2}\right)_p = 0$$

$$\left(\frac{\partial C_v}{\partial V}\right)_T = \left(\frac{\partial C_p}{\partial p}\right)_T = 0$$

$$C_p = \boxed{f(T) \text{ only (ideal gas)}}$$

$$C_V = \boxed{f(T) \text{ only (ideal gas)}}$$

10. Since at infinite dilution $x_1 = x_2 = 0$,

$$A_{12} = \ln(2) = \boxed{0.6932}$$

$$A_{21} = \ln(0.5) = \boxed{-0.6932}$$

11. In the process,

cool gas: $500°F \longrightarrow -33.4°C$ (1)
condense: $-33.4°C$ (2)
cool liquid: $-33.4°C \longrightarrow -77.7°C$ (3)
solidify: $-77.7°C$ (4)
cool solid: $-77.7°C \longrightarrow -150°F$

$$T_1 = 500°F = 533.1 \text{K}$$
$$T_2 = -33.4°C = 239.8 \text{K}$$
$$T_3 = -77.7°C = 195 \text{K}$$
$$T_4 = -150°F = 172 \text{K}$$

In completing the following steps, please refer to the specified equations from the *Chemical Engineering Reference Manual*, Fifth Edition.

(a) *step 1*: Use Eq. 4.59 for 1 gmol.

$$\Delta S_{\text{gas}} = (1)\left[6.5486 \ln\left(\frac{239.8}{533.1}\right)\right.$$
$$+ (6.1251 \times 10^{-3})(239.8 - 533.1)$$
$$+ (2.3663 \times 10^{-6})\left[\frac{(239.8)^2 - (533.1)^2}{2}\right]$$
$$\left.+ (-1.5981 \times 10^{-9})\left[\frac{(239.8)^3 - (533.1)^3}{3}\right]\right]$$

$$\Delta S_{\text{gas}} = -7.2568 \frac{\text{cal}}{\text{gmol} \cdot \text{K}}$$
$$= \frac{-7.2568}{17.032} = -0.426 \text{ cal/gram} \cdot \text{K}$$

step 2: Use Eq. 4.55.

$$\Delta S_c = \left(\frac{-5581}{239.8}\right)(1) = -23.274 \frac{\text{cal}}{\text{gmol} \cdot \text{K}}$$
$$= -1.3665 \text{ cal/gram} \cdot \text{K}$$

step 3: Use Eq. 4.58.

$$\Delta S_L = 1.06 \ln\left(\frac{195.5}{239.8}\right)$$
$$= -0.2165 \text{ cal/gram} \cdot \text{K}$$

step 4: Use Eq. 4.55.

$$\Delta S_f = \frac{-1352}{195.5}$$
$$= -6.9156 \text{ cal/gmol} \cdot \text{K}$$
$$= -0.406 \text{ cal/gram} \cdot \text{K}$$

cool solid: Use Eq. 4.58.

$$\Delta S_s = 0.502 \ln\left(\frac{172}{195.5}\right)$$
$$= -0.0643 \text{ cal/gram} \cdot \text{K}$$

$$\Sigma \Delta S = \Delta S_g + \Delta S_c + \Delta S_L + \Delta S_f + \Delta S_s$$
$$= -0.426 - 1.3665 - 0.2165 - 0.406 - 0.0643$$
$$= -2.479 \text{ cal/gram} \cdot \text{K}$$

$$= \boxed{-2.479 \text{ BTU/lbm-}°\text{R}}$$

(b) Cool gas $25°C \longrightarrow -33.4°C$

$$T_1 = 298.2 \text{K} \quad T_2 = 239.8 \text{K}$$

$$\Delta S_{\text{gas}} = (1)\left[6.5486 \ln\left(\frac{239.8}{298.2}\right) + 6.1251 \times 10^{-3}\right.$$
$$\times \left[\frac{(239.8)^2 - (298.2)^2}{2}\right] - 1.5981 \times 10^{-9}$$
$$\left.\times \left[\frac{(239.8)^3 - (298.2)^3}{3}\right]\right]$$

$$= -1.8366 \text{ cal/gmol} \cdot \text{K}$$
$$= -0.1078 \text{ cal/gram} \cdot \text{K}$$

From part (a),

$$\Delta S = -0.1078 - 1.3665 - 0.2165 - 0.4060$$
$$= -2.09868 \text{ cal/gram} \cdot \text{K}$$
$$S_{298} = 46.03 \text{ cal/gmol} \cdot \text{K}$$
$$= 2.7026 \text{ cal/gram} \cdot \text{K}$$

Since

$$\Delta S = S_{195.5} - S_{298}$$
$$S_{195.5} = -2.09868 + 2.7026$$
$$= \boxed{0.6058 \text{ BTU/lbm-}°\text{R}}$$

12.

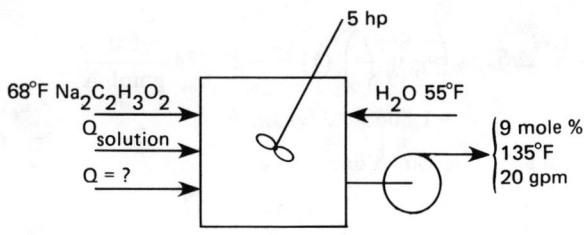

Assume the following:

- Mechanical energy ⟶ heat.
- There is no heat loss.
- The pump and mixer are 100% efficient.

$$MW\ NaC_2H_3O_2 = 82$$

For 1 lbmole,

		wgt%
H_2O: $(0.91)(18) = 16.4$ lbm	$\frac{16.4}{23.8} = 69\%$	
$Na_2C_2H_3O_2$: $(0.09)(82) = \underline{7.4}$ lbm	$\frac{7.4}{23.8} = 31\%$	
23.8 lbm		

The amount of the solution pumped is

$$(20\ \text{gpm})\left(60\ \frac{\text{min}}{\text{hr}}\right)\left(8.33\ \frac{\text{lbm}}{\text{gal}}\right)(1.15) = 11{,}500\ \text{lbm/hr}$$

The rates are

$$Na_2C_2H_3O_2\text{: } (0.31)(11{,}500) = 3565\ \text{lbm/hr}$$
$$H_2O\text{: } (0.69)(11{,}500) = 7935\ \text{lbm/hr}$$

The heat of solution is

$$\left(-3943\ \frac{\text{cal}}{\text{mole}}\right)\left(\frac{1\ \text{mole}}{82\ \text{g}}\right)\left(3.960\times 10^{-3}\ \frac{\text{BTU}}{\text{cal}}\right)\left(453\ \frac{\text{g}}{\text{lbm}}\right)$$
$$= -86.5\ \frac{\text{BTU}}{\text{lbm}}$$

Use 55°F as the datum.

The heat balance is

$$(11{,}500)(0.94)(135-55) = (3565)(0.339)(68-55)$$
$$+ (3565)(86.5) + (5+7.5)$$
$$\times \left(2544\ \frac{\text{BTU}}{\text{hr-hp}}\right) + Q$$

$$Q = \boxed{509{,}000\ \text{BTU/hr added}}$$

13.

The process is the adiabatic compression of a gas. The initial gas properties are p_1, T_1.

$$p_2 = p_1\left(\frac{T_2}{T_1}\right)^{\frac{\gamma}{(\gamma-1)}}$$

$$\gamma = 1.4\ (\text{air})$$
$$T_1 = 294\text{K}$$
$$T_2 = 573\text{K}$$
$$p_1 = 1\ \text{atm}$$
$$p_2 = (1)\left(\frac{573}{294}\right)^{\frac{1.4}{0.4}}$$
$$= \boxed{10.4\ \text{atm}}$$

14.

The weight flow is

$$(1.2\ \text{gpm})\left(\frac{1\ \text{ft}^3}{7.5\ \text{gal}}\right)\left(62.4\ \frac{\text{lbm}}{\text{ft}^3}\right) = 9.98\ \text{lbm/min}$$

(a) $h_\text{steam} = 1150.4$ BTU/lbm (tables)
$h_f = 18.07$ BTU/lbm at 50°F
$= 180.07$ BTU/lbm at 212°F
$= 117.89$ BTU/lbm at 150°F

For the balance of the change in specific enthalpy, let x represent the lbm steam.

$$(9.98)(18.07) + x(1150.4) = (9.98 + x)(117.89)$$

$$x = \boxed{0.964\ \text{lbm/min steam}}$$

(b)

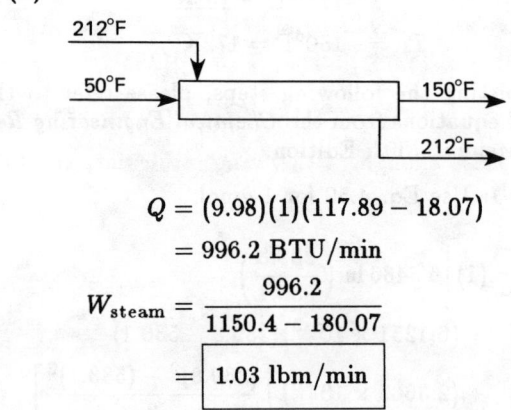

$$Q = (9.98)(1)(117.89 - 18.07)$$
$$= 996.2\ \text{BTU/min}$$
$$W_\text{steam} = \frac{996.2}{1150.4 - 180.07}$$
$$= \boxed{1.03\ \text{lbm/min}}$$

15.

	$\Delta H°$ cal
$CO + \frac{1}{2}O_2 \longrightarrow CO_2$	$-67{,}636$
$H_2 + \frac{1}{2}O_2 \longrightarrow H_2O_l$	$-68{,}313$
$H_2O_l \longrightarrow H_2O_g$	$+10{,}519$
$H_2 + CO + O_2 \longrightarrow CO_2 + H_2O$	$-125{,}430$

$$\text{moles } \frac{N_2}{O_2} = \frac{79}{21} = 3.76$$

Total moles before the reaction:

$$1_{O_2} + 1_{CO} + 1_{H_2} + 3.76_{N_2} = 6.76$$

Total moles after the reaction:

$$1_{CO_2} + 1_{H_2O} + 3.76_{N_2} = 5.76$$

$$\Delta H = \Delta U + \Delta pV$$
$$= \Delta pV$$
$$\Delta U = 0$$

For an ideal gas,

$$\Delta pV = n_2 R T_2 - n_1 R T_1$$
$$= R(n_2 T_2 - n_1 T_1)$$

$$\Delta H = \Delta H_{298} + (\bar{C}_{p_{CO_2}} + \bar{C}_{p_{H_2O}} + 3.76 \bar{C}_{p_{N_2}})(T_2 - T_1) \quad [\text{Eq. 1}]$$

$$\Delta H = R[5.76 T_2 - 6.76 T_1] \quad [\text{Eq. 2}]$$

i	C_{p_i}	$n_i C_{p_i}$ (assume 3000K)
CO_2	13.5	13.5
H_2O	11.0	11.0
N_2	8.25	31.20
		55.52

If $T_2 = 3000°K = 5400°R = 4940°F$

Combine Eq. 1 and Eq. 2.

$$-125{,}430 + (55.52)(T_2 - 298)$$
$$= (1.987)[5.76 T_2 - (6.76)(298)]$$

$$T_2 = 3131\text{K} = 2858°\text{C}$$

This is close enough. Otherwise, recalculate the table at 3100K.

$$p_2 = \frac{n_2 T_2}{n_1 T_1} p_1$$
$$= \left(\frac{5.76}{6.76}\right)\left(\frac{3131}{298}\right)(5)$$
$$= \boxed{44.8 \text{ atm}}$$

16. The airlift work is

$$W_a = (20 \text{ gpm})\left[\left(8.33 \frac{\text{lbm}}{\text{gal}}\right)\left(1.5 \frac{\text{lbm}}{\text{gal}}\right)\right](50 \text{ ft})$$
$$= 12{,}500 \text{ ft-lbm/min}$$

The isothermal expansion is

$$W_a = nRT_1 \ln\left(\frac{p_1}{p_2}\right)$$
$$= p_1 V_1 \ln\left(\frac{p_1}{p_2}\right)$$

The actual pump (airlift) work is

$$W_a = \frac{12{,}500}{0.3} = 41{,}700 \text{ ft-lbf/min}$$

$$41{,}700 = \left(14.7 \frac{\text{lbm}}{\text{in}^2}\right)\left(144 \frac{\text{in}^2}{\text{ft}^2}\right)\left[\ln\left(\frac{64.7}{14.7}\right)\right]V_1$$

$$V_1 = 13.27 \text{ ft}^3/\text{min}$$

The isentropic work for the compressor is

$$-W_c = \frac{nRT_1}{\gamma - 1}\gamma\left[\left(\frac{p_2}{p_1}\right)^{\frac{\gamma-1}{\gamma}} - 1\right]$$
$$= \left(\frac{p_1 V_1 \gamma}{\gamma - 1}\right)\left[\left(\frac{p_2}{p_1}\right)^{\frac{\gamma-1}{\gamma}} - 1\right]$$
$$= \left[\frac{(14.7)(144)(13.27)(1.4)}{0.4}\right]\left[\left(\frac{64.7}{14.7}\right)^{\frac{0.4}{1.4}} - 1\right]$$
$$= 52{,}000 \text{ ft-lbm/min}$$

The pump horsepower is

$$\frac{52{,}000}{33{,}000} = \boxed{1.57 \text{ hp}}$$

17. Assume that $K_\alpha = K_p$.

$$FeO + CO \rightarrow FeS + CO_2$$

$$K_p = \frac{y_{CO_2}}{y_{CO}} p^{(1-1)}$$
$$= \frac{y_{CO_2}}{y_{CO}} = 40$$

(a) Using a basis of 1 mole gas,

$$CO = 0.2 - x$$
$$CO_2 = x$$
$$40 = \frac{x}{0.2 - x}$$
$$x = 0.195$$
$$\text{moles Fe} = 0.195$$
$$\text{weight Fe} = (0.195)(55.8)$$
$$= \boxed{10.88 \text{ lbm}}$$

(b) Using pure CO,

$$CO = 1 - x$$
$$CO_2 = x$$
$$40 = \frac{x}{1-x}$$
$$x = 0.976$$

moles Fe = $\boxed{0.976}$

18.

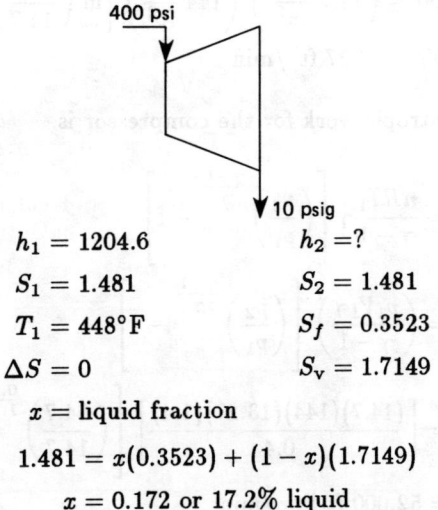

$h_1 = 1204.6$ $h_2 = ?$
$S_1 = 1.481$ $S_2 = 1.481$
$T_1 = 448°F$ $S_f = 0.3523$
$\Delta S = 0$ $S_v = 1.7149$
x = liquid fraction

$$1.481 = x(0.3523) + (1-x)(1.7149)$$
$$x = 0.172 \text{ or } 17.2\% \text{ liquid}$$

(a) For every 100 lbm of steam, 82.8 lbm of steam is available.

(b) The thermodynamic efficiency is 100%, on the basis that the heat of condensation is deducted and the result is net work.

(c) If $T_1 = 548°F$, $S_1 = 1.5541$.

$$1.5541 = x(0.3523) + (1-x)(1.7149)$$
$$x = \boxed{0.118 \text{ or } 11.8\%}$$

Thus, 88.2 lbm of process steam is available per 100 lbm of higher pressure steam.

19.

$$T_R = \frac{550 + 460}{(548)(1.8)} = 1.02$$
$$P_R = \frac{4500}{(14.7)(47.7)} = 6.4$$

For the acetonitrile, $Z = 0.83$ (from the generalized plots).

$$T_R = \frac{1010}{(126.2)(1.8)} = 4.45$$
$$P_R = \frac{10}{33.5} = 0.3$$

For the nitrogen, $Z = 1.0$.

$$m_a = \frac{pV}{ZRT} = \frac{(4500)(0.2)}{(0.83)(10.73)(1010)}$$
$$= 0.1001 \text{ lbmoles}$$
$$m_N = \frac{(10)(2)}{(1)(0.73)(1010)} = 0.0271 \text{ lbmoles}$$
$$y_a = \frac{0.1001}{0.1001 + 0.0271} = 0.787$$

Use the pseudocritical rule of Kay.

$$T_{PC} = (0.787)(548) + (0.213)(126.2)$$
$$= 458K$$
$$P_{PC} = (0.787)(47.7) + (0.213)(33.5)$$
$$= 44.7 \text{ atm}$$
$$T_R = \frac{1010}{(458)(1.8)} = 1.225$$
$$\frac{P}{P_{PC}} = \frac{ZnRT}{P_{PC}V}$$
$$= P_R = \frac{Z(0.1308)(0.73)(1010)}{(44.7)(2.2)}$$
$$P_R = Z(0.994) \text{ at } T_R = 1.225$$

Use generalized charts, at $T_R = 1.225$,

assume Z	P_r	Z (from table)
0.9	0.895	0.832
0.85	0.845	0.844
0.84	0.835	0.846

$$P_R = 0.846$$
$$P = (0.846)(44.7)$$
$$= \boxed{37.8 \text{ atm}}$$

20.

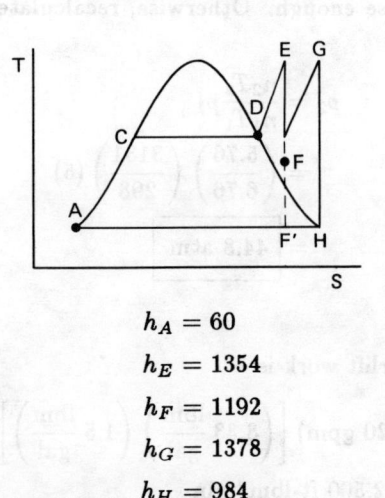

$h_A = 60$
$h_E = 1354$
$h_F = 1192$
$h_G = 1378$
$h_H = 984$
$h_{F'} = 878$

(a) $Q_1 = (h_E - h_A) + (h_G - h_F)$
$= (1354 - 60) + (1378 - 1192)$
$= 1480 \text{ BTU/lbm}$

(b) $Q_2 = h_A - h_H$
$= 60 - 984$
$= -924 \text{ BTU/lbm}$

(c) $W_{\text{net}} = Q_1 + Q_2$
$= 556 \text{ BTU/lbm}$

(d) $\eta = \dfrac{W_{\text{net}}}{Q_1}$
$= \dfrac{556}{1480} = 0.376$

(e) quality $= 100 - 11.2$
$= 88.8\%$ (Mollier chart)

In the Rankine cycle,

(a) $Q_1 = h_E - h_A = 1294 \text{ BTU/lbm}$
(b) $Q_2 = h_A - h_{F'} = -818 \text{ BTU/lbm}$
(c) $W_{\text{net}} = Q_1 + Q_2 = 476 \text{ BTU/lbm}$
(d) $\eta = \dfrac{W_{\text{net}}}{Q_1} = 0.368$
(e) quality $= 100 - 21.4 = 78.6\%$

21. At the azeotrope,

$$\gamma_i = \dfrac{p}{p_i^o}$$

$$\ln(\gamma_i) + \ln(p_i^o) = \ln(p)$$

Let $e = \left(\dfrac{A_{12}}{A_{21}}\right)\left(\dfrac{x_1}{1-x_1}\right)$

$$\ln(\gamma_1) = \dfrac{A_{12}}{(1+e)^2}$$

$$\ln(\gamma_2) = \dfrac{A_{21}}{\left(1+\dfrac{1}{e}\right)^2}$$

$$\dfrac{A_{12}}{(1+\dfrac{1}{e})^2} = \ln p - \ln p_1^o$$

$$\dfrac{A_{21}}{\left(1+\dfrac{1}{e}\right)^2} = \ln p - \ln p_2^o$$

$$\dfrac{A_{21}}{A_{12}} e^2 = K$$

$$= \dfrac{\ln p - \ln p_2^o}{\ln p - \ln p_1^o}$$

$$x_1 = \dfrac{\sqrt{\dfrac{A_{21}}{A_{12}} K}}{1 + \sqrt{\dfrac{A_{21}}{A_{12}} K}}$$

For the calculation scheme, guess t, and calculate p_1^o and p_2^o. If p_1^o and p_2^o are above 760 mm, calculate x_1.

Check that $\Sigma y_i = 1$ from

$$\dfrac{\gamma_1 x_1 p_1^o + \gamma_2 x_2 p_2^o}{p} = 1$$

	\multicolumn{6}{c}{t}
	50	55	60	65	64	63
p_1^o	612	730	863	1017	985	953
p_2^o	507	601	709	832	806	781
K				0.311	0.226	0.126
x_1				0.329	0.295	0.238
e				0.634	0.542	0.404
γ_1				0.848	0.831	0.800
γ_2				0.950	0.958	0.972
Σy_i				1.07	1.03	1001

$t = 63°\text{C}$ \quad $x_{\text{acetone}} = 0.238$

SOLUTIONS FOR CHAPTER 5
FLUID STATICS AND DYNAMICS

1. $V_{box} = 1\text{ ft}^3$
   $V_{H_2O} = 1\text{ ft}^3$
   $W_{box} = \rho_{iron} V_{iron}$
   $\rho_{iron} = (62.45)(7.2) = 449.7$
   $W_{H_2O} = \rho_{H_2O} V_{H_2O} = (62.45)(1) = 62.45\text{ lbm}$
   $V_{iron} = 1 - \left(\dfrac{11.5}{12}\right)^3 = 0.1198\text{ ft}^3$
   $W_{box} = (449.7)(0.1198) = 53.895\text{ lbm}$

   The box will float because an equal volume of box weighs less than water.

   $\text{percent submergence} = \left(\dfrac{53.895}{62.45}\right)(100)$
   $= \boxed{86\%}$

2. Since the flow is fully turbulent, f is unchanged.

   $h_L = f\left(\dfrac{L}{D}\right)\left(\dfrac{v^2}{2g}\right)$

   $\dfrac{h_1}{h_2} = \dfrac{f_1\left(\dfrac{L_1}{D_1}\right)\left(\dfrac{v_1^2}{2g}\right)}{f_2\left(\dfrac{L_1}{D_1}\right)\left(\dfrac{v_2^2}{2g}\right)}$

   $\dfrac{v_1^2}{v_2^2} = \dfrac{1}{2}$

   $v_2 = 1.4 v_1$

   Tell your manager, "The flow is increased by 40%."

3. $v = \sqrt{2g_c \Delta h}$

   $h_{air} = \dfrac{\frac{1}{12}\rho_{Hg}}{\rho_{air}}$
   $= \left(\dfrac{1}{12}\right)\left[\dfrac{(13.56)(62.4)}{0.0764}\right]$
   $= 925.7\text{ ft air}$

   $v = \sqrt{(2)(32.2)(925.7)}$
   $= \boxed{244.2\text{ ft/sec air}}$

   $h_{H_2O} = \left(\dfrac{1}{12}\right)\left[\dfrac{(13.56)(62.4)}{61.82}\right]$
   $\rho_{H_2O} = 61.862 \text{ at } 110°\text{F}$
   $= 1.143\text{ ft H}_2\text{O}$
   $v = \sqrt{(2)(32.2)(1.143)}$
   $= \boxed{8.58\text{ ft/sec H}_2\text{O}}$

4. $h_{pump} = h_{loss} + h_s + h_v + h_{outlet}$
   $h_s = 100\text{ ft}$
   $h_v = 0$
   $h_{loss} = 0.0311 f\left(\dfrac{LQ^2}{d^5}\right)$

   $v = (0.408)\left(\dfrac{Q}{d^2}\right)$
   $= (0.408)\left[\dfrac{750}{(6.065)^2}\right]$
   $= 8.319\text{ ft/sec}$

   $N_{Re} = \dfrac{Dv\rho}{\mu_e}$
   $= \dfrac{\left(\dfrac{6.065}{12}\right)(8.319)\left[(0.9)(62.45)\right]}{0.0015}$
   $= 1.57 \times 10^5$
   $f = 0.0185$

   From Darcy's formula in Crane's "Flow of Fluids Through Valves and Fittings,"

   $h_{loss} = 0.0311 f\left(\dfrac{LQ^2}{d^5}\right)$
   $= (0.0185)(0.0311)\left[\dfrac{3000}{(6.065)^5}\right](750)^2$
   $= 118\text{ ft}$

   $h_{outlet} = \dfrac{(50)(144)}{(0.9)(62.45)} = 128$
   $h_{pump} = 118 + 100 + 0 + 128$
   $= 346\text{ ft oil}$
   $p = \dfrac{(346)(0.9)(62.45)}{144}$
   $= \boxed{135\text{ psi}}$

5. $\dfrac{L}{D} = R_T + (n-1)\left(R_L + \dfrac{R_B}{2}\right)$

From the formula for resistance of bends in Crane's "Flow of Fluids Through Valves and Fittings," ninth printing,

$$\dfrac{r}{d} = \dfrac{3}{0.21} = 14.3$$

$$\dfrac{L}{D} = 39 + (47)\left(22 + \dfrac{17}{2}\right)$$

$$= \boxed{1472.5 \text{ diameters}}$$

6. $\dfrac{h_1}{h_2} = \dfrac{\dfrac{64}{N_{\text{Re 1}}}}{\dfrac{64}{N_{\text{Re 2}}}}$

$= \dfrac{N_{\text{Re 2}}}{N_{\text{Re 1}}} = \dfrac{\mu_1}{\mu_2}$

$\dfrac{h_1}{h_2} = \dfrac{1}{0.5} = 2$

$h_2 = 0.5 h_1$

The pressure drop decreases by one-half.

7. See appendix B, "Flow of Water Through Schedule 40 Steel Pipe," in Crane's.

At 20 gpm, in a 1-inch schedule-40 pipe,

$$\Delta p = \dfrac{10.9 \text{ psi}}{100 \text{ ft}}$$

$$= (10.9)\left(\dfrac{65}{100}\right)$$

$$= \boxed{7.09 \text{ psi}}$$

8. Refer to appendix B, "Flow of Air Through Schedule 40 Steel Pipe," in Crane's.

At 100 ft^3/min,

$$\Delta p = 0.534 \text{ psi}/100 \text{ ft}$$

The ideal gas correction is

$$\Delta p = (0.8)(0.534)\left(\dfrac{100 + 14.7}{112 + 14.7}\right)\left(\dfrac{460 + 95}{520}\right)$$

$$= \boxed{0.418 \text{ psi}}$$

9. See Table 5.11, "Equivalent Length of Straight Pipe for Various Fittings."

$$\text{ells: } (3)(13) = 39 \text{ ft}$$
$$\text{couplings: } (4)(0.65) = 2.6 \text{ ft}$$

Refer to appendix A, "Equivalent Lengths L and L/D and Resistance Coefficient K," in Crane's.

The gate valve is half-open.

$$\dfrac{L}{D} = 160$$

$$L = (160)\left(\dfrac{4.026}{12}\right)$$

$$= 53.7 \text{ ft}$$

Refer to "Dimensionless Groupings and Their Significance," from Perry's *Chemical Engineers' Handbook*.

plug cock $\Theta = 20°$

$$\dfrac{L}{D} = (1.56)(45)$$

$$L = (1.56)(45)\left(\dfrac{4.026}{12}\right) = 23.6$$

exchanger: $L = (200)\left(\dfrac{4.026}{12}\right)$

$$= 67.1 \text{ ft}$$

$L_{\text{total}} = 400 + 67.1 + 39 + 2.6 + 53.7 + 23.6$

$L = 586.0 \text{ ft}$

See appendix B, "Flow of Water Through Schedule 40 Steel Pipe," in Crane's.

$$\Delta p = 5.65 \text{ psi}/100 \text{ ft}$$

$$= (5.65)\left(\dfrac{586.0}{100}\right)$$

$$= \boxed{33.1 \text{ psi}}$$

10.

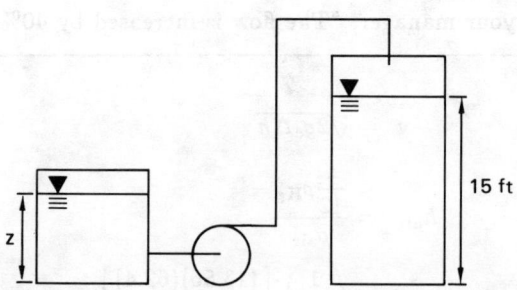

See appendix B, "Flow of Water Through Schedule 40 Steel Pipe," in Crane's.

For schedule-40 2-inch pipe,

$$1 \dfrac{\text{ft}}{\text{sec}} = 10.45 \text{ gpm}$$

$$\text{I.D.} = 2.067 \text{ in}$$

$$A = 0.0233 \text{ ft}^2$$

Using Bernoulli's equation,

$$z + h_{pump} = 15 + \frac{v^2}{2g} + h_f$$

$$h_{pump} = \left(\frac{1}{8}\text{ hp}\right)\left(550\,\frac{\text{ft-lbf}}{\text{hp-sec}}\right)\left(\frac{1\text{ min}}{5\text{ ft}^3}\right)\left(\frac{1\text{ ft}^3}{62.4\text{ lbf}}\right)\left(60\,\frac{\text{sec}}{\text{min}}\right)$$

$$= 13.22\text{ ft}$$

$$h_f = 0.8\text{ ft}$$

$$v = \left(5\,\frac{\text{ft}^3}{\text{min}}\right)\left(7.48\,\frac{\text{gal}}{\text{ft}^3}\right)\left(\frac{1\,\frac{\text{ft}}{\text{sec}}}{10\,\frac{\text{gal}}{\text{min}}}\right)$$

$$= 3.57\text{ ft/sec}$$

$$z = -13.22 + 15 + \left[\frac{(3.57)^2}{(2)(32.2)}\right] + 0.8$$

$$z = \boxed{2.8\text{ ft}}$$

11. Refer to appendix A, "Representative Resistance Coefficients (K) for Valves and Fittings," in Crane's.

$$\text{relative radius} = \frac{\frac{72}{2}}{3} = 12$$

The equivalent length of one 90° bend is 34.5 pipe diameters.

R_t = total resistance one 90° bend
n = total number 90° bends
R_L = resistance due to length
R_B = resistance due to bend

$$\frac{L}{D} = R_t + (n-1)\left(R_L + \frac{R_B}{2}\right)$$

$$R_L = 18.7$$
$$R_B = 15.2$$
$$h = 60$$

$$\frac{L}{D} = 34.5 + (60-1)\left(18.7 + \frac{15.2}{2}\right)$$

$$= 1600\text{ diameters}$$

$$L = (1600)\left[\frac{3 - (2)(0.065)}{12}\right]$$

$$= 384\text{ ft}$$

$$N_{Re} = (50.6)\left(\frac{Q\rho}{d\mu}\right)$$

$$Q = 50\text{ gpm}$$
$$d = 2.87\text{ in}$$
$$\rho = 62.4\text{ lbm/ft}^3$$
$$\mu = 0.862\text{ cp}$$

$$N_{Re} = (50.6)\left[\frac{(150)(62.4)}{(2.87)(0.862)}\right] = 190{,}000$$

$$f = 0.0195\text{ (Moody chart)}$$

$$\Delta p = 0.000216\left(\frac{fL\rho Q^2}{d^5}\right)$$

$$= (0.000216)(0.0195)\left[\frac{(384)(62.4)(150)^2}{(2.87)^5}\right]$$

$$= \boxed{11.7\text{ psi}}$$

12.
$$N_{Re} = (123.9)\left(\frac{dv\rho}{\mu}\right)$$

$$= (123.9)\left[\frac{(1)(7.33)(55)}{25}\right]$$

$$= 2000\text{ laminar}$$

$$f = \frac{64}{N_{Re}} = 0.032$$

The new product is

$$N_{Re} = (2000)\left(\frac{25}{12.5}\right) = 4000$$

$$f_{new} = 0.0392\text{ (Moody chart)}$$

The increase in horsepower proportional to f is

$$\text{percent increase hp} = \left(\frac{0.0392 - 0.032}{0.032}\right)(100)$$

$$= \boxed{22.5\%}$$

13. $W = 450{,}000 + 50{,}000 = 500{,}000\text{ lbm/hr}$

$$\text{solution: }(100)\left(\frac{450{,}000}{500{,}000}\right) = 90\%$$

$$90\%\text{ solution SG} = 1.2$$
$$10\%\text{ crystals SG} = 2.1$$

$$\text{SG slurry} = \frac{1}{\frac{0.9}{1.2} + \frac{0.1}{2.1}} = 1.254$$

$$C_v = 400 = Q\sqrt{\frac{SG}{\Delta p}}$$

Using Bernoulli's equation,

$$z_1 + \frac{p_1}{\rho} = h_{pump} + h_f + h_{valve} + z_2 + \frac{p_2}{\rho}$$

$$Q = \left(5 \times 10^5\,\frac{\text{lbf}}{\text{hr}}\right)\left[\frac{1\text{ ft}^3}{(62.4)(1.254)\text{ lbf}}\right]$$

$$\times \left(\frac{1\text{ hr}}{60\text{ min}}\right)\left(7.48\,\frac{\text{gal}}{\text{ft}^3}\right)$$

$$= 797\text{ gal/min}$$

$$C_v = 400 = 797\sqrt{\frac{1.254}{\Delta p}}$$

$\Delta p = 4.98$ psi across valve

$$\frac{\Delta p}{\rho} = h_{\text{valve}} = \frac{(4.98)(144)}{(62.4)(1.254)}$$

$$= 9.15 \text{ ft}$$

$$\frac{p_1 - p_2}{\rho} = h_A$$

$$= -26 \text{ in Hg}$$

$$= (-26)\left(\frac{13.5}{1.253}\right)\left(\frac{1}{12}\right)$$

$$= -23.3 \text{ ft}$$

$$h_A = h_{\text{valve}} + h_{\text{pump}} + h_{\text{friction}} + h_{\text{elevation}}$$

$$-h_{\text{pump}} = 23.3 + 9.15 + 18 + 30$$

$$= 80.45 \text{ ft slurry}$$

$$= (80.45)(1.254)$$

$$= \boxed{100.9 \text{ ft H}_2\text{O}}$$

The sign is negative for the pump work on the fluid.

14. Using Bernoulli's equation,

$$z_1 + \frac{v_1^2}{2g} + \frac{p_1}{\rho} = z_2 + \frac{v_2^2}{2g} + \frac{p_2}{\rho} + h_f$$

$$z_1 - \frac{v_2^2}{2g} - h_f = 0$$

Since $p_1 - p_2 = 0$,

$$h_f = f\left(\frac{L}{D}\right)\left(\frac{v^2}{2g}\right)$$

$$z_1 - \left(\frac{v_2^2}{2g}\right)\left[1 + f\left(\frac{L}{D}\right)\right] = 0$$

$$v_2 = \sqrt{\frac{2gz_1}{1 + f\left(\frac{L}{D}\right)}}$$

$$Q = v_2 A$$

$A = 0.0233 \text{ ft}^2$

$$Q = v_2(0.0233)\left(7.48 \frac{\text{gal}}{\text{ft}^3}\right)\left(60 \frac{\text{sec}}{\text{min}}\right)$$

$$= 10.46 v_2$$

$$= \boxed{10.46 \sqrt{\frac{2gz_1}{1 + f\left(\frac{L}{D}\right)}} \text{ gpm}}$$

15.
$$N_{\text{Re}} = 123.9 \frac{dv\rho}{\mu}$$

$$= (123.9)\left[\frac{(4.026)(10)(62.4)}{4.35}\right]$$

$$= 71{,}600$$

$$f = 0.0188$$

$$h_f = f\left(\frac{L}{D}\right)\left(\frac{v^2}{2g}\right)$$

$$= (0.0188)\left[\frac{(5280)(5)(12)}{4.026}\right]\left[\frac{(10)^2}{(2)(32.2)}\right]$$

$$= 2297 \text{ ft}$$

$$\text{heat} = (2297 \text{ ft})\left(1.285 \times 10^{-3} \frac{\text{BTU}}{\text{ft-lbm}}\right)$$

$$= 2.95 \text{ BTU/lbm}$$

$$= C_p \Delta T$$

$$\Delta T = \frac{\text{heat}}{C_p} = \frac{2.95 \frac{\text{BTU}}{\text{lbm}}}{0.2 \frac{\text{BTU}}{\text{lbm-}^\circ\text{F}}}$$

$$= \boxed{14.8^\circ\text{F temperature rise}}$$

16. The pressure drop due to friction loss is

$$h_f = f\left(\frac{L}{D}\right)\left(\frac{v^2}{2g}\right)$$

Gas properties are a function of temperature. Assume

$$\Delta p = 16 \text{ psi}$$

$$p_{\text{average}} = \frac{60 + (60 + 16)}{2} = 68 \text{ psig}$$

$$= 82.7 \text{ psia}$$

At 82.7 psia, 70°F,

$$\rho_{\text{CH}_4} = 0.22 \frac{\text{lbm}}{\text{ft}^3}$$

$$\mu = 7.39 \times 10^{-6} \frac{\text{lbm}}{\text{ft-sec}}$$

$$v = \frac{Q}{A}$$

$$Q = (250)\left(\frac{14.7}{82.7}\right)$$

$$= 44.4 \text{ ft}^3/\text{sec}$$

$$A = \pi \frac{I^2}{4} = 0.786 \text{ ft}^2$$

$$v = \frac{44.4}{0.786} = 56.58 \text{ ft/sec}$$

$$N_{\text{Re}} = \frac{Dv\rho}{\mu} = \frac{(1)(56.58)(0.22)}{7.39 \times 10^{-6}}$$

$$= 1.684 \times 10^6$$

FLUID STATICS AND DYNAMICS

Using the Colebrook and White equation,

$$f = \left[1.8 \log\left(\frac{N_{Re}}{7}\right)\right]^{-2} = 0.0106 \text{ smooth pipe}$$

$$\Delta p = 0.001294 \left(\frac{fL\rho v^2}{d}\right)$$

$$= (0.001294)(0.0106)\left[\frac{(3)(5280)(0.22)(56.58)^2}{12}\right]$$

$$= 12.8 \text{ psi}$$

If $\epsilon/D = 0.00010$, using the Sacham equation,

$$f = \left\{-2\log\left[\frac{\frac{\epsilon}{d}}{3.7} - \frac{5.02}{N_{Re}}\log\left(\frac{\frac{\epsilon}{d}}{3.7} + \frac{14.5}{N_{Re}}\right)\right]\right\}^{-2}$$

$$= 0.0129$$

$$\Delta p = (0.001294)\left[\frac{(0.0129)[(3)(5280)](0.22)(56.58)^2}{12}\right]$$

$$= 15.6 \text{ psi (close enough)}$$

The reservoir pressure is $60 + 15.6 = \boxed{75.6 \text{ psig}}$

17. (a) The net pump head is

$$\Sigma_{\text{elevation}} + \Sigma_{\text{pressure}} + \Sigma_{\text{velocity}}$$

$$\Sigma_{\text{elevation}} = 4 \text{ ft}$$

$$\Sigma_{\text{pressure head}} = \left[30.7 - \left(14.7 + \left(\frac{5}{29.92}\right)(14.7)\right)\right](2.3039)$$

$$= 31.2 \text{ ft}$$

$$\Sigma_{\text{velocity head}} = \frac{v_2^2 - v_1^2}{2g}$$

$$v_2 = (818 \text{ gpm})\left[\frac{0.408}{(5)^2}\right]$$

$$= 13.3 \text{ ft/sec}$$

$$v_1 = v_2 \left(\frac{5}{10}\right)^2$$

$$= 3.33 \text{ ft/sec}$$

$$\Sigma_{\text{velocity head}} = \frac{(13.3)^2 - (3.33)^2}{2g}$$

$$= 2.57 \text{ ft}$$

The net pump head is

$$4 + 31.2 + 2.57 = 37.8 \text{ ft}$$

$$\text{hydraulic hp} = \frac{(\text{gpm})(\text{head})(8.33)}{33,000}$$

$$= \frac{(818)(37.7)(8.33)}{33,000} = 7.78 \text{ hp}$$

$$\text{efficiency \%} = 100 \frac{\text{hydraulic hp}}{\text{input hp}}$$

$$= (100)\left(\frac{7.8}{10}\right)$$

$$= \boxed{78\%}$$

(b) If the pump speed is increased to 3500,

$$Q_2 = Q_1 \left(\frac{N_2}{N_1}\right) = (818)\left(\frac{3500}{1750}\right)$$

$$= 1636 \text{ gpm}$$

$$h_1 = h_2 \left(\frac{N_2}{N_1}\right)^2 = (37.8)\left(\frac{3500}{1750}\right)^2$$

$$= 151.2 \text{ ft}$$

$$W_2 = W_1 \left(\frac{N_2}{N_1}\right)^3 = (10)\left(\frac{3500}{1750}\right)^3$$

$$= 80 \text{ hp}$$

18. The pressure drop of compressible fluids is

$$\Delta p = 3.36 \times 10^{-6} \frac{fW^2 L}{d^5 \rho}$$

L/D for elbows $= 30$

The equivalent length is

$$23 + (30)\left(\frac{6}{12}\right) = 68 \text{ ft}$$

$$\Delta p = 100 \text{ psi}$$
$$d^5 = 6346 \text{ in}^5$$
$$\mu = 0.021$$
$$d = 5.76$$
$$\rho = \frac{(MW)p}{RT}$$
$$= \frac{(28)(114.7)}{[(10.72)(212 + 460)]}$$
$$= 0.4458 \text{ lbm/ft}^3$$

Rearranging,

$$W = \sqrt{\frac{\Delta p d^5 \rho}{3.36 \times 10^{-6} f L}}$$

$$= \sqrt{\frac{(100)(6346)(0.4458)}{[(3.36 \times 10^{-6} f)(68)]}}$$

$$= 35{,}188 \left(\sqrt{\frac{1}{f}}\right)$$

$$N_{\text{Re}} = \frac{6.31 W}{d \mu}$$

$$= \frac{6.31 W}{(5.761)(0.021)}$$

$$= 52.15 W$$

$$f = \left\{-2\log\left[\frac{\frac{\epsilon}{D}}{3.7} - \frac{5.02}{N_{\text{Re}}} \log\left(\frac{\frac{\epsilon}{D}}{3.7} - \frac{14.5}{N_{\text{Re}}}\right)\right]\right\}^{-2}$$

For steel pipe, use $\epsilon = 0.002$.

$$\frac{\epsilon}{D} = \frac{0.002}{\frac{5.761}{12}} = 0.004$$

Using the Moody chart (Fig. 5.8), if $N_{\text{Re}} > 2.5 \times 10^5$ for $\epsilon/D = 0.004$, then f is constant.

By trial and error,

- Guess f.
- Calculate W.
- Calculate N_{Re}.
- Calculate f.
- Compare f_{guess} to $f_{\text{calculated}}$.

f_{guess}	W	N_{Re}	$f_{\text{calculated}}$
0.0288	2.073×10^5	1.081×10^7	0.0284
0.0284	2.087×10^5	1.089×10^7	0.0284

The maximum CO_2 rate is 208,700 lbm/hr.

19. Using the Bernoulli equation,

$$z_1 + \frac{p_1}{\rho} + \frac{v_1^2}{2g} = z_2 + \frac{p_2}{\rho} + \frac{v_2^2}{2g} + h_f + h_{\text{pump}}$$

$$z_1 = 0$$
$$p_1 = p_2$$
$$v_1 = 0$$
$$z_2 = 4000 \text{ ft}$$
$$v_2 = 10 \frac{\text{ft}}{\text{sec}}$$

$$h_{\text{pump}} = -\left(z_2 + \frac{v_2^2}{2g} + h_f\right)$$

$$\frac{v_2^2}{2g} = \frac{100}{(2)(32.2)} = 1.6 \text{ ft}$$

$$Q = \frac{v d^2}{0.408} = \frac{(10)(5.761)^2}{0.408}$$
$$= 813 \text{ gpm}$$

$$\mu = 1.129 C_p$$

$$N_{\text{Re}} = 123.9 \left(\frac{d v \rho}{\mu}\right)$$

$$= (123.9)(5.761)\left[\frac{(10)(62.4)}{1.129}\right]$$

$$= 394{,}511$$

Assume $\epsilon/D = 0.0003$. Using the Sacham equation,

$$f = \left\{-2\log\left[\left(\frac{\frac{\epsilon}{d}}{3.7} - \frac{5.02}{N_{\text{Re}}}\log\right)\left(\frac{\frac{\epsilon}{d}}{3.7} + \frac{14.5}{N_{\text{Re}}}\right)\right]\right\}^{-2}$$

$$= 0.0163$$

$$h_f = f\left(\frac{L}{D}\right)\left(\frac{v^2}{2g}\right)$$

$$= (0.0163)\left(\frac{5000}{\frac{5.761}{12}}\right)(1.6)$$

$$= 271 \text{ ft}$$

$$h_{\text{pump}} = -(4000 + 1.6 + 271)$$
$$= -4273 \text{ ft}$$

$$\text{pump hp} = \frac{Q h_{\text{pump}} (8.33)}{33{,}000}$$

$$= \frac{(813)(4273)(8.33)}{33{,}000} = 877 \text{ hp}$$

$$\text{Bhp} = \frac{876}{0.7} = 1252 \text{ hp}$$

$$\text{cost} = (1252)\left(0.745 \frac{\text{kW}}{\text{hp}}\right)\left(\frac{\$0.015}{\text{kW-hr}}\right)$$

$$= \boxed{\$14.00/\text{hr}}$$

20. For spherical particles, $\psi = 1$.

$$\epsilon = 1.08 - [(1.12)(1)] + [(0.405)(1)^2]$$
$$= 0.365$$

$$V_{\text{bed}} = \left(\frac{\pi D^2}{4}\right)(L)$$

$$= \left(\frac{\pi 4^2}{4}\right)(6) = 75.4 \text{ ft}^3$$

$$A = 12.6 \text{ ft}^2$$

flow $= (20)(75.4) = 1508$ gpm

$$v_s = (1508)(0.1337)\left(\frac{1}{60}\right)\left(\frac{1}{12.6}\right)$$

$\quad = 0.2667$ ft/sec

$\rho = 62.4$ lbm/ft^3

$\mu = (1)(6.72 \times 10^{-4})$

$\quad = 6.72 \times 10^{-4}$ lbm/ft-sec

$$N_{\text{Re}}^* = \frac{(62.4)\left(\dfrac{0.065}{12}\right)(0.2667)}{6.72 \times 10^{-4}}$$

$\quad = 134.1$

Using the Ergun equation,

$$\Delta p = \left[\frac{(1-0.365)(62.4)(0.2667)^2(6)}{(0.365)^3\left(\dfrac{0.065}{12}\right)(32.2)}\right]$$

$$\times \left[(150)\left(\frac{1-0.365}{134.1}\right) + 1.75\right]$$

$\Delta p = (1993.8)(2.46) = 4905$ lbm/ft^2

$\quad = \boxed{34 \text{ psi}}$

SOLUTIONS FOR CHAPTER 6

HEAT TRANSFER: CONDUCTION AND RADIATION

1. $$\frac{q}{A} = \frac{t_1 - t_s}{\Sigma \frac{l_i}{k_i} + \frac{1}{h}} = 200$$

    $$200 = \frac{1400 - 85}{\left[\left(\frac{1}{12}\right)\left(\frac{9}{0.8} + \frac{4.5}{0.15} + \frac{9}{0.4}\right)\right] + \frac{1}{h}}$$

    $$h = \boxed{0.792 \text{ BTU/hr-ft}^2\text{-}°F}$$

2. The minimum wall thickness of the fireclay brick is

    $$\frac{q}{A} = 250 = \frac{\Delta t}{\frac{L}{k}} = \frac{2000 - 1800}{\frac{L}{0.9}}$$

    $L = 0.72 \text{ ft} = 8.64 \text{ in} = 2 \text{ bricks}$

    $$\frac{q}{A} = 250 = \frac{t_1 - t_2}{\left(\frac{1}{12}\right)\left(\frac{9}{0.9}\right)} = \frac{2000 - t_2}{0.833}$$

    $t_2 = 1791.7°F$

    For the insulating brick,

    $$250 = \frac{1791.7 - 300}{\frac{L_2}{k_2}} = \frac{1491.7}{\frac{L_2}{0.12}}$$

    $L_2 = 0.716 \text{ ft} = 8.59 \text{ in} = 3 \text{ bricks}$

    $$250 = \frac{1791.7 - t_3}{\left(\frac{1}{12}\right)\left(\frac{9}{0.12}\right)}$$

    $t_3 = 229.2°F$

    For the building brick,

    $$250 = \frac{229.2 - 100}{\frac{L_3}{k_3}} = \frac{129.2}{\frac{L_3}{0.4}}$$

    $L_3 = 0.207 \text{ ft} = 1 \text{ brick}$

    (a) The total thickness is

    $$9 + 9 + 4 = \boxed{22 \text{ in}}$$

    (2 fireclay + 3 insulating + 1 building brick)

(b) $$\frac{q}{A} = \frac{2000 - 100}{\left(\frac{1}{12}\right)\left(\frac{9}{0.9} + \frac{9}{0.12} + \frac{4}{0.4}\right)}$$

    $$= \boxed{240 \text{ BTU/hr-ft}^2}$$

3. $$q = \frac{4\pi k \Delta T r_i r_o}{r_i - r_o}$$

    $r_i = \frac{3}{12} \text{ ft}$

    $r_o = \frac{5}{12} \text{ ft}$

    $$q = \frac{(4\pi)(8)(212 - 300)\left(\frac{3}{12}\right)\left(\frac{5}{12}\right)}{\frac{3}{12} - \frac{5}{12}}$$

    $$= \boxed{5529 \text{ BTU/hr}}$$

4. Ignore edge and corner losses. For thickness x,

    $$q = \frac{\Delta t}{R}$$

    $$R = \frac{x}{kA}$$

    $$A = (2)\Big[(5)(4) + (5)(3) + (4)(3)\Big]$$

    $$= 94 \text{ ft}^2$$

    But

    $$q = \frac{1{,}440{,}000}{24} = 60{,}000 \text{ BTU/hr}$$

    $$R = \frac{500 - 100}{60{,}000}$$

    $$= 0.665 \times 10^{-2} \text{ hr-}°F/\text{BTU}$$

    $$x = RkA = (0.665 \times 10^{-2})(0.8)(94)$$

    $$= 0.5 \text{ ft} = \boxed{6 \text{ in}}$$

5.
$$q = \frac{T_1 - T_3}{\left(\frac{1}{2\pi L}\right)\left[\frac{\ln\left(\frac{r_2}{r_1}\right)}{k_1} + \frac{\ln\left(\frac{r_3}{r_2}\right)}{k_2}\right]}$$

$k_{\text{asbestos}} = 0.12$ BTU/hr-ft-°F
$k_{\text{cork}} = 0.03$ BTU/hr-ft-°F
$r_1 = \frac{1}{12}$ ft
$r_2 = \frac{3}{12}$ ft
$r_3 = \frac{4.5}{12}$ ft

$$q = \frac{300 - 85}{\left(\frac{1}{2\pi}\right)(1)\left[\frac{\ln\left(\frac{3}{1}\right)}{0.12} + \frac{\ln\left(\frac{4.5}{3}\right)}{0.03}\right]}$$

$$= \boxed{59.59 \text{ BTU/hr-ft of pipe}}$$

6. Heat transfer takes place by radiation and natural convection.

$$q_r = A\sigma(T_1^4 - T_2^4)\left(\frac{1}{\frac{1}{\epsilon_1} + \frac{1}{\epsilon_2} - 1}\right)$$

$$= (150)(0.1713 \times 10^{-8})[(530)^4 - (460)^4]$$

$$\times \left(\frac{1}{\frac{1}{0.9} + \frac{1}{0.9} - 1}\right) = 7180$$

$$q_c = h_c A(T_1 - T_2)$$

$$\frac{h_c x}{k_f} = \left[\left(\frac{x_p f^2 g \beta_f \Delta t}{\mu_f^2}\right)\left(\frac{C_p \mu}{k}\right)\right]^m \left[\frac{C_*}{\left(\frac{L}{x}\right)^{\frac{1}{9}}}\right]$$

(a) Assume no aluminum and $\epsilon = 0.9$ for wood. The air properties at 35°F are

$C_p = 0.240$ BTU/lbm-°F
$\mu = 0.042$ lbm/ft-hr
$k = 0.0141$ BTU/hr-ft-°F
$\frac{C_p \mu}{k} = 0.72 = N_{\text{Pr}}$

Let

$$Y = \frac{\rho_f^2 \beta_f g C_p}{\mu_f k_f} = 2.16 \times 10^6 \text{ ft-°F}$$

$C_* = 0.071$ and $m = \frac{1}{3}$ when $N_{\text{Gr}} = 2.1 \times 10^5$ to 1.1×10^7.

$$\frac{\frac{1}{3} h_c}{0.0141} = \left[\frac{0.071}{\left(\frac{10}{\frac{1}{3}}\right)^{\frac{1}{9}}}\right](Y\Delta t x^3)^{\frac{1}{3}}$$

$$= \left(\frac{0.071}{(30)^{\frac{1}{9}}}\right)\left[\frac{(2.16 \times 10^6)(70)}{27}\right]^{\frac{1}{3}}$$

$h_c = 0.3645$
$q_c = (0.3645)(150)(70) = 3828$ BTU/hr
$q = q_r + q_c = 7180 + 3828$

$$= \boxed{11,008 \text{ BTU/hr}}$$

(b) Assume that aluminum is present. Since the properties of air are not much different on either side of the foil,

$\epsilon_{\text{Al}} = 0.087$
$q_r = (150)(0.1713 \times 10^{-8})[(530)^4 - (495)^4]$

$$\times \left(\frac{1}{\frac{1}{9} + \frac{1}{0.087} - 1}\right)$$

$= 418$ BTU/hr

$$N_{\text{Gr}} = \frac{Y x^3 \Delta t}{N_{\text{Pr}}} = \frac{(1.92 \times 10^6)\left(\frac{1}{6}\right)^3 (35)}{0.72}$$

$= 4.32 \times 10^5$
$c = 0.071$
$m = \frac{1}{3}$

$$\frac{h_c x}{k_f} = \left(\frac{0.071}{(60)^{\frac{1}{9}}}\right)[(4.32 \times 10^5)(0.72)]^{\frac{1}{3}}$$

$h_c = 0.265$
$q_c = (150)(0.267)(35) = 1402$
$q = h_c + h_r$
$= 1402 + 418 = 1820$ BTU/hr

$$\frac{q_{\text{Al}}}{q_{\text{without}}} = \frac{1820}{11,008} = 0.165$$

The heat transfer is reduced by 84% by using aluminum.

7. $Y = \dfrac{t' - t}{t' - t_b} = \dfrac{60 - 100}{60 - 300} = 0.1666$

$n = \dfrac{r}{r_m} = \dfrac{0}{\frac{1}{12}} = 0$

$m = \dfrac{k}{r_m h} = 0$

Since $h \gg k/r_m$, X is determined from Fig. 6.7.

$X = 0.83 = \dfrac{\alpha \Theta}{r_m^2}$

$\alpha = \dfrac{k}{\rho C_p} = \dfrac{0.4}{(155)(0.2)} = 0.0129$

$\Theta = (0.83)\left(\dfrac{r_m^2}{\alpha}\right)$

$= \dfrac{(0.83)\left(\dfrac{1}{12}\right)^2}{0.0129}$

$= \boxed{0.446 \text{ hr}}$

HEAT TRANSFER: CONVECTION AND EQUIPMENT

1.

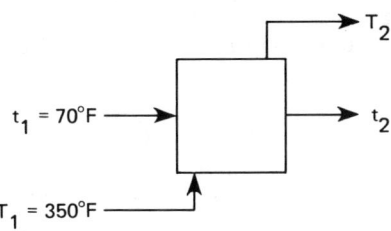

The heat balance is

$$Q = wC_p(T_1 - T_2)$$
$$= wC_p(t_2 - t_1)$$
$$(10{,}000)(1)(350 - T_2) = (10{,}000)(1)(t_2 - 70)$$
$$350 - T_2 = t_2 - 70$$

Since

$$T_2 - t_2 = 5°F$$
$$2t_2 = 415°F$$
$$t_2 = 207.5°F$$
$$Q = (10{,}000)(1)(207.5 - 70)$$
$$= \boxed{1.375 \times 10^6 \text{ BTU/hr}}$$

2. $\Delta t_{lm} = 10°F$

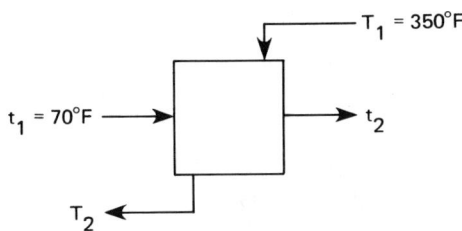

Since the flows and C_p are the same, Δt is the same at both ends.

$$10 = 350 - t_2 = T_2 - 70$$
$$t_2 = 340$$
$$Q = (10{,}000)(1)(340 - 70)$$
$$= \boxed{2.7 \times 10^6 \text{ BTU/hr}}$$

3. On the F_T chart,

$$P_{\max} = 0.58 \quad \left(R = \frac{w_c}{w_c} = 1.0\right)$$

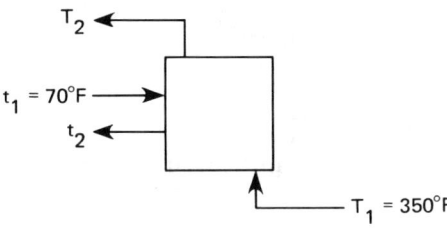

$$P = \frac{t_2 - 70}{T_1 - 70} = 0.58$$
$$= \frac{t_2 - 70}{350 - 70}$$
$$t_2 = 232°F$$
$$Q = (10{,}000)(1)(232 - 70)$$
$$= \boxed{1.62 \times 10^6 \text{ BTU/hr}}$$

4.

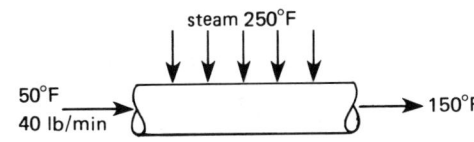

$\frac{3}{4}$ in BWG tube: 0.065 wall
0.1623 ft²/ft

$$a = 0.3019 \text{ in}^2 \text{ flow area}$$
$$D_i = 0.620 \text{ in}$$
$$C = 471 \text{ lbm-hr/ft-sec}$$
brass: $k = 770$ BTU/hr-ft²-°F/in

The water velocity is

$$v = \frac{(40)(60)}{(471)(1)} = 5.1 \text{ ft/sec}$$
$$t_{\text{avg}} = \frac{50 + 150}{2} = 100°F$$

From Fig. 7-10,

$$h_i = \boxed{1230 \text{ BTU/hr-ft}^2\text{-°F}}$$

5. $U_i = \dfrac{1}{\dfrac{1}{h_i} + \left(\dfrac{L}{k}\right)\left(\dfrac{D_i}{D_m}\right) + \left(\dfrac{1}{h_o}\right)\left(\dfrac{D_i}{D_o}\right)}$

$= \dfrac{1}{\dfrac{1}{1230} + \left(\dfrac{0.065}{770}\right)\left(\dfrac{0.620}{0.685}\right) + \left(\dfrac{1}{1000}\right)\left(\dfrac{0.620}{0.750}\right)}$

$= \boxed{583 \text{ BTU/hr-ft}^2\text{-}°\text{F}}$

6. $\Delta t_m = \dfrac{200 - 100}{\ln\left(\dfrac{200}{100}\right)} = 144.3°\text{F}$

$A_i = \dfrac{Q}{U_i \Delta t_m}$

$= \dfrac{(40)(60)(1)(150 - 50)}{(583)(144.3)}$

$= 2.85 \text{ ft}^2$

$L = \dfrac{2.85}{0.1623}$

$= \boxed{17.6 \text{ ft}}$

7. $U_D = \dfrac{1}{2} U_i$

or $U_D = \left(\dfrac{1}{2}\right)(583)$

$= 291.5 \text{ BTU/hr-ft}^2\text{-}°\text{F}$

$U_D = \dfrac{1}{\dfrac{1}{U_i} + \dfrac{1}{h_\sigma}}$

$= 291.5 = \dfrac{1}{\dfrac{1}{583} + \dfrac{1}{h_\sigma}}$

$h_\sigma = \boxed{583 \text{ BTU/hr-ft}^2\text{-}°\text{F}}$

8. The solution to this problem is based on the assumptions stated.

$q_{\text{in}} = (q_{\text{out}})_{\text{radiation}} + (q_{\text{out}})_{\text{convection}}$

Assumption 1: Radiation is lost to the air at 90°F.

$\epsilon = 1$

$350 = (0.173)(1)\left[\left(\dfrac{t_s + 460}{100}\right)^4 - \left(\dfrac{550}{100}\right)^4\right] + (2.8)(t_s - 90)$

Using the bisection method,

$\boxed{t_s = 172.3°\text{F}}$

Assumption 2: Radiation is lost to space at 0°R.

$\epsilon = 1$

$350 = (0.173)(1)\left[\left(\dfrac{t_s + 460}{100}\right)^4 - 0\right] + (2.8)(t_s - 90)$

$t_s = \boxed{136°\text{F}}$

Assumption 3: There is no radiation loss.

$350 = (2.8)(t_s - 90)$

$t_s = \boxed{215°\text{F}}$

9. The enthalpy balance is

(a) $w_{H_2O} C_p (t_o - t_i) = \Delta h_{\text{vapor}} w_{CH_3OH}$

$w_{H_2O}(1)(115 - 70) = (600)(10{,}000)$

$w_{H_2O} = \boxed{133{,}333 \text{ lbm/hr}}$

(b) $q = UA \Delta t_{\text{lm}}$

$\Delta t_{\text{lm}} = \dfrac{(160 - 70) - (160 - 115)}{\ln\left(\dfrac{160 - 70}{160 - 115}\right)} = 65°\text{F}$

$q = \Delta H_{\text{vapor}} w_{CH_3OH}$

$= (600)(10{,}000)$

$= 6 \times 10^6 \text{ BTU/hr}$

$6 \times 10^6 = (400) A (65)$

$A = \boxed{231 \text{ ft}^2}$

(c)
- There is no heat loss to the surroundings.
- U is constant.
- C_p is constant.
- The flow is steady.

HEAT TRANSFER: CONVECTION AND EQUIPMENT

10.

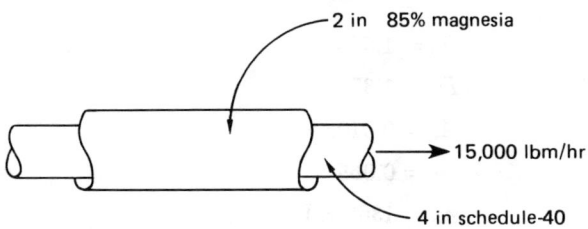

$U_{\text{surface}} = 1.75 \text{ BTU/hr-ft}^2\text{-}°\text{F}$

The air temperature is 80°F.

$$A = (75 \text{ ft})(\pi)\left(\frac{8.5}{12}\right) = 167 \text{ ft}^2$$
$$Q = UA\Delta t = (1.75)(167)(100 - 80)$$
$$= 5841 \text{ BTU/hr}$$

The amount of vapor condensed is

$$\frac{5841 \dfrac{\text{BTU}}{\text{hr}}}{55 \dfrac{\text{BTU}}{\text{lbm}}} = 106.2 \text{ lbm/hr}$$

$$\text{quality} = \frac{15{,}000 - 106.5}{15{,}000}$$
$$= \boxed{99.3\%}$$

11.

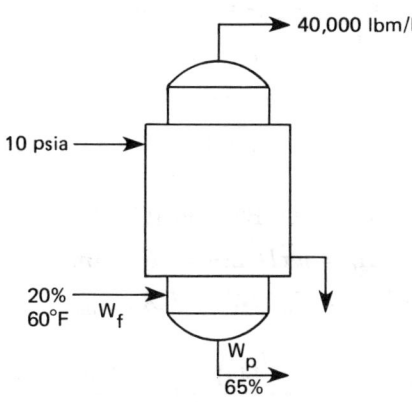

$p = 4 \text{ in Hg}$
$h = (125.4 - 32)(0.93) = 86.8$
$t = 125.4°\text{F} = t_{\text{sat}}$
$h_v = 1116.0 \text{ BTU/lbm}$
$h_{\text{feed}} = C_p \Delta t$
$= (60 - 32)(0.93) = 26 \text{ BTU/lbm}$

The solids balance is

$$W_f(0.2) = (0.65)W_p$$

The overall balance is

$$W_p + 40{,}000 = W_f$$
$$W_p = 17{,}750$$
$$W_f = 57{,}750$$
$$\text{temp in evaporator} = 125.4°\text{F}$$

The enthalpy balance is

$$Q = (86.8)(17{,}750) + (40{,}000)(1116.0) - (26)(57{,}750)$$
$$= 4.467 \times 10^7 \text{ BTU/hr} = UA\Delta t$$

$$A = \frac{Q}{U\Delta t}$$
$$= \frac{4.467 \times 10^7}{(250)(193.2 - 125.4)}$$
$$= \boxed{2636 \text{ ft}^2}$$

The amount of steam required is

$$\frac{Q}{\Delta h_v} = S$$
$$S = \frac{4.4679 \times 10^7}{982.1}$$
$$= \boxed{45{,}550 \text{ lbm/hr steam}}$$

12. (a) An approximate calculation can show the effect of oil inside versus oil outside. For the oil outside,

$$\frac{1}{U_o} \approx \frac{1}{h_o} + \frac{1}{h_i} = \frac{1}{41} + \frac{1}{2000} = 0.0249 \text{ hr-ft}^2\text{-}°\text{F/BTU}$$

For the oil inside,

$$\frac{1}{U_o} \approx \frac{1}{h_o} + \frac{1}{h_i} = \frac{1}{90} + \frac{1}{840} = 0.0123 \text{ hr-ft}^2\text{-}°\text{F/BTU}$$

U_o is about twice as large for oil on the inside. Since oil is the controlling fluid, it should be on the inside where U is the highest.

(b) For one-inch schedule-40 pipe,

$$D_i = 1.049 \text{ in}$$
$$D_o = 1.315 \text{ in}$$
$$\pi D_o = 0.344 \text{ ft}$$
$$\text{wall} = X_W = 0.133 \text{ in}$$
$$k_m = 26 \text{ BTU-ft/hr-ft}^2\text{-}°\text{F}$$

$$\frac{1}{U_o} = \frac{1}{h_o} + \frac{X_W D_o}{k_m D_m} + \frac{D_o}{h_i D_i}$$
$$= \frac{1}{840} + \left[\frac{(0.133)(1.315)}{(12)(1.18)}\right]\left(\frac{1}{26}\right) + \frac{1.315}{(90)(1.049)}$$

$$U_o = 64 \text{ BTU/hr-ft}^2\text{-}°\text{F}$$

$$q = UA\Delta t$$
$$= WC_p(t_i - t_o)$$
$$= (8000)(0.4)(150 - 40)(1.8)$$
$$= 633{,}600 \text{ BTU/hr}$$
$$\Delta t = \frac{(150 - 40) - (40 - 20)}{\ln\left(\frac{110}{20}\right)} = 52.8^\circ\text{C}$$

$$U_o A\Delta t = U\pi D_o L\Delta t = q$$
$$L = \frac{q}{U\pi D_o \Delta t}$$
$$= \frac{633{,}600}{(64)(0.344)(52.8)(1.8)}$$
$$= \boxed{303 \text{ ft}}$$

13. Assume that the air film controlling $h = U$, and 1 means first case. From the Sieder-Tate equation it can be shown that if other fluid properties remain constant,

$$U_1 = kV_1^{0.8}$$
$$U_2 = kV_2^{0.8}$$
$$q_1 = W_1 C_p(t_2 - t_1) = U_1 A\Delta t_1$$
$$q_2 = W_2 C_p(t_2 - t_1) = U_2 A\Delta t_2$$
$$\Delta t_1 = \frac{(220 - 80) - (220 - 180)}{\ln\left(\frac{140}{40}\right)} = 79.8^\circ\text{F}$$
$$\Delta t_2 = \frac{(250 - 80) - (250 - 170)}{\ln\left(\frac{170}{70}\right)} = 113^\circ\text{F}$$

W is proportional to the velocity.

$$\frac{W_2}{W_1} = \frac{W_2^{0.8} \Delta t_2}{W_1^{0.8} \Delta t_1}$$

or

$$\frac{W_2^{0.2}}{W_1^{0.2}} = \frac{\Delta t_2}{\Delta t_1} = 1.416$$
$$W_2 = W_1(1.416)^5$$
$$= W_1(5.7)$$

$\boxed{\text{Approximately 5.7 times as much air can be heated with higher temperature steam.}}$

14. From the tubing characteristics,

$$D_o = 1.5 \text{ in}$$
$$D_i = 1.37 \text{ in}$$
$$A_i = 0.0102 \text{ ft}^2$$
$$x_W = 0.065$$
$$h_o = 1300 \text{ BTU/hr-ft}^2\text{-}^\circ\text{F}$$
$$h_i = 600 \text{ BTU/hr-ft}^2\text{-}^\circ\text{F}$$
$$k_m = 105 \text{ BTU-in/hr-ft}^2\text{-}^\circ\text{F}$$

Assume the design is for 6 ft/sec.

$$\mathbf{v} = 6 \text{ ft/sec}$$
$$\rho = (1.3)(62.4) = 81.2 \text{ lbm/ft}^3$$
$$W = 650{,}000 \text{ lbm/hr}$$
$$m = \text{number of passes}$$
$$n = \text{number of tubes}$$

$$nA_i\bar{v}\rho = W$$
$$n = \frac{650{,}000}{(0.0102)(6)(3600)(81.12)}$$
$$= 36.4 = 36 \text{ tubes}$$
$$\frac{1}{U_o} = \frac{1}{h_o} + \frac{D_o X_W}{\bar{D}k_m} + \frac{D_o}{h_i D_i}$$
$$\bar{D} = \frac{1.5 - 1.37}{\ln\left(\frac{1.5}{1.37}\right)} = 1.43 \text{ in}$$
$$\frac{1}{U_o} = \frac{1}{1300} + \frac{(1.5)(0.065)}{(1.43)(105)} + \frac{1.5}{(1.37)(600)}$$

$$U_o = 308 \text{ BTU/hr-ft}^2\text{-}^\circ\text{F}$$
$$A_o = n\pi D_o Lm = (169.7)m$$
$$q = C_p \Delta t W = U_o A_o \Delta t_{lm}$$

$$\Delta t_{lm} = \frac{(t_s - 148) - (t_s - 198)}{\ln\left(\frac{t_s - 148}{t_s - 198}\right)}$$
$$t_s = 258.8^\circ\text{F}$$
$$\Delta t_{lm} = 83.5^\circ\text{F}$$
$$q = (0.78)(198 - 148)(650{,}000)$$
$$= (308)(169.7m)(83.5)$$
$$m = 5.82$$

$\boxed{\text{Use 6 passes of 36 tubes per pass.}}$

15. For the concurrent heat exchanger,

$$Q_{\text{oil}} = m_{\text{oil}} C_p (t_o - t_i)$$
$$= (2000)(0.56)(200 - 60)$$
$$= 123{,}200 \text{ BTU/hr}$$
$$Q_{\text{kerosene}} = 123{,}200 \text{ BTU/hr}$$
$$m_{\text{kerosene}} = \frac{123{,}200}{(0.6)(450 - 220)}$$
$$= 893 \text{ lbm/hr}$$

For the countercurrent heat exchanger,

$$123{,}200 = m(0.6)(450 - 110)$$
$$m = 604 \text{ lbm/hr}$$

> The flow rate is greater for the concurrent heat exchanger.

Compare at the same kerosene flow, 893 lbm/hr.

$$(\Delta t_{\text{lm}})_{\text{concurrent}} = \frac{360 - 20}{\ln\left(\frac{360}{20}\right)} = 118°\text{F}$$

$$(\Delta t_{\text{lm}})_{\text{counter}} = \frac{250 - 130}{\ln\left(\frac{250}{130}\right)} = 183°\text{F}$$

Where Q and U are fixed, the countercurrent exchanger requires less area at the same flow rate.

At minimum flow rates,

$$(\Delta t_{\text{lm}})_{\text{counter}} = \frac{250 - 20}{\ln\left(\frac{250}{20}\right)} = 91.2°\text{F}$$

> At minimum flow rates the countercurrent heat exchanger needs more area.

16. (a)
$$\frac{1}{U_1} = \frac{1}{U_c} + \frac{t}{k} + \frac{1}{U_{l_1}}$$
$$\frac{1}{250} = \frac{1}{1200} + \frac{0.12}{460} + \frac{1}{U_{l_1}}$$
$$U_{l_1} = 344$$

Area of $1\frac{1}{2}$ in tubing,

having 0.12 wall = 0.00865 ft^2
and having 0.065 wall = 0.0102 ft^2

$$V_2 = V_1 \left(\frac{0.00865}{0.0102}\right) = 0.845 V_1$$

$$U_{l_2} = U_{l_1} \sqrt{\frac{0.845 V_1}{V_1}} = 316$$
$$\frac{1}{U_2} = \frac{1}{1200} + \frac{0.065}{105} + \frac{1}{316}$$

$$\boxed{217 \frac{\text{BTU}}{\text{hr-ft}^2\text{-°F}} = U_2}$$

(b) > A PCU is the amount of heat required to raise 1 lbm of H$_2$O one degree Centigrade.

$$1 \text{ PCU} = 1.8 \text{ BTU}$$

17.
$$q = U A \Delta t_{\text{lm}}$$
$$\Delta t_{\text{lm}} = \frac{(80.1 - 35) - (80.1 - 65)}{\ln\left(\frac{45.1}{15.1}\right)}$$
$$= 27.5°\text{C}$$
$$= 49.5°\text{F}$$
$$A_o = (36)(18)(0.2618) = 169.5 \text{ ft}^2$$
$$q = (5000)(170) = 8.5 \times 10^5 \text{ BTU/hr}$$

$$U_o = \frac{q}{A_o \Delta t_{\text{lm}}} = \frac{8.5 \times 10^5}{(169.5)(49.5)}$$
$$= 101 \text{ BTU/hr-ft}^2\text{-°F}$$

$$\frac{1}{U_o} = \frac{D_o}{D_i h_i} + \frac{x_W D_o}{k_m \bar{D}_i} + \frac{1}{h_o}$$
$$\frac{1}{101} = \frac{1.0}{0.87 h_i} + \frac{(0.065)(1.0)}{(12)(9.4)(0.94)} + \frac{1}{300}$$
$$h_i = 193 \text{ BTU/hr-ft}^2\text{-°F}$$

For heat transfer in the tubes,

$$h_i = 0.023 \frac{k}{D_i} (N_{\text{Re}})^{0.8} (N_{\text{Pr}})^{\frac{1}{3}}$$

$$\left(\frac{C_p \mu}{k}\right)^{\frac{1}{3}} = \left[\frac{(1.0)(0.56)(2.42)}{0.372}\right]^{\frac{1}{3}}$$
$$= 1.54$$

$$\text{H}_2\text{O rate} = \frac{q}{C_p \Delta t} = \frac{8.5 \times 10^5}{(1.0)(30)(1.8)}$$
$$= 1.57 \times 10^4 \frac{\text{lbm}}{\text{hr}}$$

In the cross section,

$$(36)(0.00413) = 0.1485 \text{ ft}^2$$
$$G = \frac{1.57 \times 10^4}{0.1485}$$
$$= 1.06 \times 10^5 \text{ lbm/hr-ft}^2$$

$$N_{\text{Re}} = \frac{DG}{\mu} = \frac{\left(\frac{0.87}{12}\right)(1.06 \times 10^5)}{(0.56)(2.42)}$$

$$= 5650$$

$$N_{\text{Re}}^{0.8} = 1000$$

$$h_i = \frac{(0.023)(0.372)(1000)(1.54)}{\frac{0.87}{12}}$$

$$= 182 \text{ BTU/hr-ft}^2\text{-}^\circ\text{F}$$

Since h_i was actually 193, the heat exchanger is performing properly with no evidence of fouling.

18. The enthalpy balance is

	aniline	toluene
t_{avg}	125°F	
C_p	0.52	0.46

For the aniline,

$$Q = wc_p \Delta t$$
$$= (7000)(0.52)(150 - 100)$$
$$= 182{,}000 \text{ BTU/hr}$$

For the toluene,

$$\Delta t = \frac{Q}{WC_p}$$
$$= \frac{182{,}000}{(10{,}000)(0.46)}$$
$$= 39.5^\circ\text{F}$$

$$T_{\text{out}} = 185 - 39.5 = 145.5^\circ\text{F}$$

$$\Delta t_{\text{lm}} = \frac{45.5 - 35}{\ln\left(\frac{45.5}{35}\right)} = 40^\circ\text{F}$$

Using the following equations,

$$D_e \text{ annulus} = D_2 - D_1$$
$$G = \frac{W}{A}$$
$$N_{\text{Re}} = \frac{DG}{\mu}$$
$$N_{\text{Pr}} = \frac{C_p \mu}{k}$$
$$h = 0.023 \frac{k}{D} N_{\text{Re}}^{0.8} N_{\text{Pr}}^{\frac{1}{3}}$$

	hot: annulus, toluene	cold: inner aniline
$D_e =$	0.625 ft	0.0874 ft
$G =$	719,000 lbm/ft²-hr	1,162,000 lbm/ft²-hr
$\mu =$	0.848 lbm/ft-hr	4.72 lbm/ft-hr
$N_{\text{Re}} =$	5.3×10^4	2.51×10^4
$N_{\text{Pr}} =$	4.63	25
$h_o =$	340	176

$$\frac{1}{U_o} = \frac{D_o}{D_i h_i} + \frac{x_W D_o}{k_m D_L} + \frac{1}{h_o} + \frac{1}{h_{\text{dirty}}}$$

$$\frac{x_W D_p}{k_m D_L} \approx \left(\frac{1}{26}\right)\left(\frac{0.133}{12}\right)\left(\frac{1.315}{1.182}\right)$$

$$= 0.47 \times 10^{-3}$$

$$\frac{1}{U_o} = \frac{1}{176} + (0.47 \times 10^{-3}) + \frac{1}{340} + (5 \times 10^{-3})$$

$$= 0.014092 \text{ hr-ft}^2\text{-}^\circ\text{F/BTU}$$

$$U_o = 71$$

$$Q = U_o A_o \Delta t$$

$$A_o = \frac{182{,}000}{(71)(40)} = 64.1 \text{ ft}^2$$

$$L = \frac{64.1 \text{ ft}^2}{0.344 \frac{\text{ft}^2}{\text{ft}}} = 186 \text{ ft}$$

Each hairpin section is 30 feet long. Therefore,

$$\frac{186}{30} = \boxed{6 \text{ hairpins}}$$

SOLUTIONS FOR CHAPTER 8
VAPOR-LIQUID PROCESSES

1. $\Sigma p_i = \Sigma p_i^o x_i = 3000$
$= 14{,}000 x_E + 2700 x_p + 500 x_B$

Since $x_p = x_B$,

$x_E + x_p + x_B = 1$

$$x_p = x_B = \frac{1 - x_E}{2}$$

$$3000 = 14{,}000 x_E + \left(\frac{1 - x_E}{2}\right)(2700 + 500)$$

$x_E = \boxed{0.1129}$

$x_p = \boxed{0.4435}$

$x_B = \boxed{0.4435}$

$$y_i = \frac{p_i}{p} = \frac{p_i^o x_i}{p}$$

$$y_E = \frac{(14{,}000)(0.1129)}{3000} = \boxed{0.5269}$$

$$y_p = \frac{(2700)(0.4435)}{3000} = \boxed{0.3992}$$

$$y_B = \frac{(500)(0.4435)}{3000} = \boxed{0.0739}$$

2. The minimum H_2O rate occurs when the bottoms are in equilibrium with the feed.

$MW_{air} = 29$
$MW_{acetone} = 58$
$MW_{mixture} = (0.02)(58) + (0.98)(29) = 29.58$

acetone in feed $= (0.02)\left(\dfrac{1000}{29.58}\right)$
$= 0.676$ lbmole/hr

acetone in bottoms $= (0.95)(0.676)$
$= 0.642$ lbmole/hr

From equilibrium,

$M_A = 2.53 x_a = 2.0$ mole %

$x_a = \dfrac{2}{2.53} = 0.791$ mole %

$x_{H_2O} = 100 - 0.791$
$= 99.209$ mole %

H_2O in bottoms $= \dfrac{(99.209)(0.642)}{0.791}$
$= 80.52$ lbmole/hr

$(80.52)(18) = \boxed{1449 \text{ lbm } H_2O/\text{hr}}$

3. Assume that equimolarity occurs in the gas phase and that A and B are organic acids.

$x_A p_A^o + x_B p_B^o = 200 - p_{H_2O}$

$y_A = y_B$

$x_A + x_B = 1$

$x_A p_A^o = x_B p_B^o = (1 - x_A) p_B^o$

$$x_A = \frac{p_A^o}{p_A^o + p_B^o} = \frac{32}{32 + 14}$$

$x_A = 0.696$

$x_B = 0.304$

$(0.696)(14) + (0.304)(32) = 200 - p_{H_2O}$

$p_{H_2O} = 180.52$

$$y_{H_2O} = \frac{180.52}{200} = 0.9026$$

$$y_A = y_B = \frac{(0.696)(14)}{200}$$
$= 0.0487$

$$\frac{\text{moles } H_2O}{\text{mole acid}} = \frac{0.9026}{(2)(0.0487)}$$

$$\frac{\text{lbm } H_2O}{\text{mole acid}} = \left[\frac{0.9026}{(2)(0.0487)}\right](18)$$

$= \boxed{167 \text{ lbm } H_2O/\text{mole acid}}$

4. (a) molal diffusivity $= D_m = D_\rho$
$= 0.577 \times 10^{-3}$ lbmole/ft-hr

For an ideal gas,

$$\rho = \frac{p}{RT}$$

$$D = \frac{D_m RT}{p}$$

$$N_A = \left(\frac{p}{RT}\right) D \left[\frac{\Delta p_A}{(\Delta z)(p_B)_{lm}}\right]$$

$$= D_m \left[\frac{\Delta p_A}{(\Delta z)(P_B)_{lm}}\right]$$

$$\Delta p_A = \frac{20}{760} - 0 = 0.026315 \text{ atm}$$

$$p_{B_1} = 1 - 0.026315 = 0.97368 \text{ atm}$$
$$p_{B_2} = 1 \text{ atm}$$
$$(p_b)_{lm} = \frac{1 - 0.97368}{\ln\left(\frac{1}{0.97368}\right)}$$
$$= 0.9866 \text{ atm}$$
$$\Delta z = 5 \text{ ft}$$
$$N_A = \frac{(0.577 \times 10^{-3})(0.026315)}{(5)(0.9866)}$$
$$= 3.078 \times 10^{-6} \text{ lbmole/ft}^2\text{-hr}$$
$$N_A A = \left(\frac{\pi}{4}\right)\left(\frac{2}{12}\right)^2 (3.078 \times 10^{-6})$$
$$= 6.715 \times 10^{-8} \text{ lbmole/hr}$$

1% volume ≡ 0.01 mole fraction
$$\ln\left(\frac{1-0.01}{1-0.026315}\right) = \left(\frac{h}{\Delta z}\right)\left[\ln\left(\frac{1-0}{1-0.026315}\right)\right]$$
$$h = 3.115$$
$$5 - h = \boxed{1.885 \text{ ft from end of pipe}}$$

5. Use a basis of 1 mole feed.
$$x_i = \frac{z_i}{k_i V + L}$$

i	z_i	k_i	x_i	y_i
CH_4	0.10	17.0	0.0111	0.1887
C_2H_6	0.20	3.1	0.0976	0.3026
C_3H_3	0.30	1.0	0.3000	0.3000
iC_4H_{10}	0.15	0.44	0.2083	0.0917
mC_4H_{10}	0.20	0.32	0.3030	0.0970
mC_5H_{12}	0.05	0.0986	0.0910	0.0090
			1.0110	0.9890

$$\boxed{\text{The equilibrium constants are consistent.}}$$

6. Use a basis of 1 mole feed.
$$F = D + W = 1$$

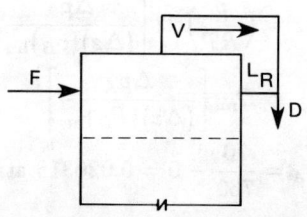

$$x_f F = x_D D + x_W W$$
$$(0.2)(1) = x_D D + (0.03)W$$
$$x_D D = (0.9)(0.2)(1) = 0.18$$
$$W = 0.67$$
$$D = 0.33$$
$$x_D = 0.546$$
$$V = L_R + D = (1 + R)D$$
$$R = 2 \text{ and } 5$$

With liquid feed, if $R = 2$,
$$V = (1+2)(0.33) = 1$$
$$\text{slope} = \frac{L}{V} = \frac{\left(\frac{5}{3}\right)}{1} = 1.667$$
$$L = RD + 1$$
$$= \frac{2}{3} + 1 = 1.667$$

If $R = 5$,
$$V = (1+5)(0.33) = \frac{6}{3}$$
$$L = RD + 1 = \frac{8}{3}$$
$$\text{slope} = \frac{L}{V} = 1.333$$

From the McCabe-Thiele plot,

For $R = 2$, assume 2.5 theoretical plates.
$$\text{actual} = \frac{2.5 - 1}{0.4} = 3.75 + \text{reboiler}$$

For $R = 5$, assume 2.3 theoretical plates.
$$\text{actual} = \frac{2.3 - 1}{0.4} = 3.25 + \text{reboiler}$$

Assume:
- For $R = 2$, 4 plates are needed.
- For $R = 5$, 3 plates are needed.

The feed is
$$\frac{60}{3600} = 0.0167 \text{ moles/sec}$$
$$t_{\text{average}} = 237°\text{F}$$
$$V = (359)\left(\frac{697}{460}\right)$$
$$= 509 \text{ ft}^3/\text{mole}$$

VAPOR-LIQUID PROCESSES

For $R = 2$,

$$V = 0.0167$$
$$= 8.50 \text{ ft}^3/\text{sec}$$
$$\text{minimum area} = \frac{8.5}{3.0} = 2.83 \text{ ft}^2$$
$$D = \sqrt{\frac{2.83}{0.7854}} = 1.89 \approx 2 \text{ ft}$$
$$\text{cost} = (300)(4) = \$1200.00$$

$$\text{steam } \$ = (0.0167)\left(\frac{14{,}500}{100}\right)[(3600)(24)(365)](\$0.0005)$$
$$= \$35{,}200.00/\text{year}$$

$$\text{water } \$ = (0.0167)\left(\frac{14{,}500}{100}\right)(3.15 \times 10^6)\left[\frac{\$0.12}{(1000)(8.33)}\right]$$
$$= \$110.00$$

Similarly, for $R = 5$,

$$D \approx 3 \text{ ft}$$
$$V = 0.0332$$
$$\text{cost} = (350)(3) = \$1050.00$$
$$\text{steam} = (35{,}200)\left(\frac{0.0332}{0.0167}\right) = \$70{,}400$$
$$\text{water} = (110)\left(\frac{0.0332}{0.0167}\right) = \$220$$

	$R = 2$	$R = 5$
column	$ 1200	$ 1050
steam	35,200	70,400
water	110	220
	$36,510	$71,670

Use $R = 2$ because it is less expensive.

7.

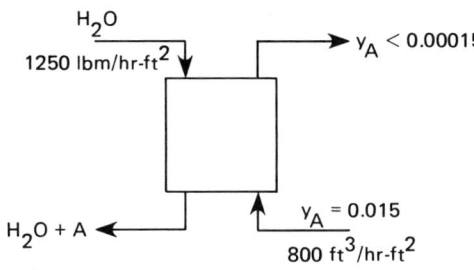

$$y = 1.75x$$
$$K_G a = 1.82 \text{ lbmoles/ft}^3\text{-hr}$$
$$G = \frac{800}{359} = 2.23 \text{ lbmoles/hr-ft}^2$$

$$L = \frac{1250}{18} = 69.5 \text{ lbmoles/hr-ft}^2$$

$$(\text{NTU})_{OG} = \left(\frac{A}{A-1}\right) \ln\left(\frac{1 - \frac{E}{A}}{1 - E}\right)$$

$$E = \frac{0.015 - 0.00015}{0.015} = 0.99$$

$$A = \frac{69.5}{(1.75)(2.23)} = 17.809$$

(a) $(\text{NTU})_{OG} = \left(\frac{17.809}{16.809}\right) \ln\left(\frac{1 - \frac{0.99}{17.809}}{1 - 0.99}\right)$

$$= \boxed{4.818}$$

(b) $(\text{HTU})_{OG} = \frac{G}{K_Y a} = \frac{2.23}{1.82} = \boxed{1.223 \text{ ft}}$

(c) $z = (\text{HTU})_{OG}(\text{NTU})_{OG}$

$$= (1.223)(4.818) = \boxed{5.89 \text{ ft}}$$

8. Assume that H_2O condenses first.

(a) and (b) $p^o_{H_2O} = 0.3 \text{ atm} = 4.41 \text{ psia}$

From the steam tables at $156.96°F$ (or $616.6°R$), (Note: 6 = hexane; 7 = heptane)

$$y_6 = p^o_6 x_6$$
$$y_6 = 0.6$$
$$y_7 = 0.1$$
$$y_s = 0.3$$
$$\ln p^o_6 = 17.7109 - \frac{6816.4}{616.6} = 6.6566$$
$$p^o_6 = 778 \text{ mm Hg}$$
$$\ln p^o_7 = 17.9184 - \frac{7547.4}{616.6} = 5.6786$$
$$p^o_7 = 292.2 \text{ mm Hg}$$
$$x_6 + x_7 = \frac{(0.6)(760)}{778} + \frac{(0.1)(760)}{292.6}$$
$$= 0.845$$
$$\Sigma x_i < 1$$

Therefore, no condensation of hydrocarbons occurs. At $156.96°F$ pure water condenses.

At $t = 150°F$ (or $609.7°R$),

$$p^o_s = 0.2530 \text{ atm} = 192.28 \text{ mm Hg}$$
$$y_6 + y_7 = \frac{760 - 192.28}{760} = 0.747$$

Since $y_6/y_7 = 6$,

$$y_6 = 0.6402$$
$$y_7 = 0.1067$$
$$\ln p_6^o = 6.5304$$
$$p_6^o = 685.7 \text{ mm Hg}$$
$$\ln p_7^o = 5.5389$$
$$p_7^o = 254.4 \text{ mm Hg}$$
$$x_6 + x_7 = \frac{(0.6402)(760)}{685.7} + \frac{(0.1067)(760)}{254.4}$$
$$= 1.028$$

Since
$$x_6 + x_7 = 0.7096 + 0.3188 = 1.028$$

(c) By interpolation, $\boxed{t = 151.1°\text{F}}$.

(d) $\quad x_6 = \dfrac{0.7096}{1.028} = \boxed{0.690}$

$\quad\quad x_7 = \dfrac{0.3188}{1.028} = \boxed{0.3108}$

SOLUTIONS FOR CHAPTER 9

DISTILLATION, EVAPORATION, AND HUMIDIFICATION

1. This problem is worded the same as problem 3 in Chapter 8. However, this problem assumes that equimolar acids occur in the liquid phase.

$$p_A = x_A p_A^o$$
$$x_C = 0.05$$
$$x_A = x_B = 0.475$$
$$\pi = \Sigma p_i$$
$$= (0.475)(32) + (0.475)(14) + p_{H_2O} = 200$$
$$p_{H_2O} = 178.15 \text{ mm Hg}$$
$$y_{H_2O} = \frac{178.15}{200} = 0.891$$
$$y_A + y_B = 1 - 0.891 = 0.109$$
$$\frac{\text{lbm H}_2\text{O}}{\text{mole acid}} = \frac{(0.891)(18)}{0.109}$$
$$= \boxed{146.76 \text{ lbm H}_2\text{O/mole acid}}$$

2. $537 \text{ kg} = 1183.9 \text{ lbm}$
    $\text{MW C}_7\text{H}_{16} = 100$
    $\text{MW C}_8\text{H}_{18} = 114$
    $$l_i = (1183.9)(0.5)\left(\frac{1}{100} + \frac{1}{114}\right)$$
    $$= 11.111 \text{ moles}$$

Note: $7 = \text{C}_7\text{H}_{16}$, $8 = \text{C}_8\text{H}_{18}$

(a) $$x_7 = \frac{(1183.9)(0.5)(0.01)}{(1183.9)(0.5)\left(0.01 + \frac{1}{114}\right)}$$
$$= 0.5327$$
$$-\ln\left(\frac{l_f}{l_i}\right) = \left(\frac{1}{\alpha - 1}\right)\left[\ln\left(\frac{x_i}{x_f}\right) - \alpha\ln\left(\frac{1 - x_i}{1 - x_f}\right)\right]$$
$$\ln\left(\frac{11.111}{4.74}\right) = (1)\left[\ln\left(\frac{0.5327}{x_f}\right) - 2\ln\left(\frac{0.4673}{1 - x_f}\right)\right]$$

This reduces to
$$0.961 = \frac{(1 - x_f)^2}{x_f}$$
$$x_f^2 - 2.961x_f + 1 = 0$$
$$x_f = \boxed{0.389} = x_7$$
$$x_8 = 1 - 0.389 = \boxed{0.611}$$

(b) $$y_7 = \frac{\alpha x}{1 + (\alpha - 1)x}$$
$$= \frac{(2)(0.5327)}{1 + (1)(0.5327)} = \boxed{0.6951}$$

(c) $$y_8 = 1 - 0.6951 = \boxed{0.3049}$$

When $\alpha = 1$, no enrichment takes place.

$$\boxed{x_7 = 0.5327} \quad \boxed{x_8 = 0.4673}$$

3. (a) Using the Fenske equation,
$$N_m = \frac{\ln\left[\left(\frac{0.9}{0.1}\right)\left(\frac{0.9}{0.1}\right)\right]}{\ln 2} - 1$$
$$= \boxed{5.34 \text{ or 6 plates}}$$

(b) The slope of the operating line at minimum reflux is calculated from the line through $(0.9, 0.9)$ and $(0.5, y^*)$.

$$y^* = \frac{\alpha x_f}{1 + (\alpha - 1)x_f} = 0.667$$
$$\text{slope} = S = \frac{0.9 - 0.667}{0.9 - 0.5} = 0.58325$$
$$S = \frac{R_m}{1 + R_m} \quad \text{or} \quad R_m = \frac{S}{1 - S}$$
$$= \frac{0.58325}{1 - 0.58325}$$
$$R_m = \boxed{1.399}$$

(c) $R = (1.2)(1.399) = 1.679$
$$y = \left(\frac{1.679}{1 + 1.679}\right)x + \frac{0.9}{1 + 1.679}$$
$$y_D = 0.627x + 0.336$$

Using the Underwood equation,
$$0.627K^2 + (0.627 + 0.336 - 2)K + 0.336 = 0$$
$$0.627K^2 - 1.037K + 0.336 = 0$$
$$K^2 - 1.654K + 0.536 = 0$$
$$K_1 = 0.442 \quad K_2 = 1.212$$

Since the feed line is vertical, it has the equation $x = x_f$. The intersection of the rectifying operating line and feed line is (x_f, y_i).

$$y_i = (0.627)(0.5) + 0.336 = 0.650$$

The stripping line slope is

$$\left(\frac{L}{V}\right)_s = \frac{0.65 - 0.1}{0.5 - 0.1} = 1.375$$

The stripping operating line is

$$y = 1.375x - 0.0375$$

Using the Underwood equation for stripping,

$$1.375K^2 + (1.375 - 0.0375 - 2)K - 0.0375 = 0$$
$$1.375K^2 - 0.6625K - 0.0375 = 0$$
$$K^2 - 0.4818K - 0.027 = 0$$

$$K_1 = -0.0507 \qquad K_2 = 0.5325$$

Rectifying,

$$n_R \ln\left(\frac{\frac{2}{0.627}}{[1 + (1)(0.442)]^2}\right)$$

$$= \ln\left[\frac{(0.9 - 0.442)(1.212 - 0.5)}{(0.5 - 0.442)(1.212 - 0.9)}\right]$$

$$0.428 n_R = 2.892$$

$$n_R = \boxed{6.756 \text{ stages}}$$

Stripping,

$$n_S \ln\left[\frac{\frac{2}{1.375}}{[1 + (1)(-0.0507)]^2}\right]$$

$$= \ln\left[\frac{(0.5 + 0.0507)(0.5325 - 0.1)}{(0.1 + 0.0507)(0.5325 - 0.5)}\right]$$

$$0.4788 n_S = 4.6687$$

$$n_S = \boxed{9.75 \text{ stages}}$$

The minimum number of theoretical stages is 20: 7 rectifying and 13 stripping stages.

4. Refer to the graphical solution provided.

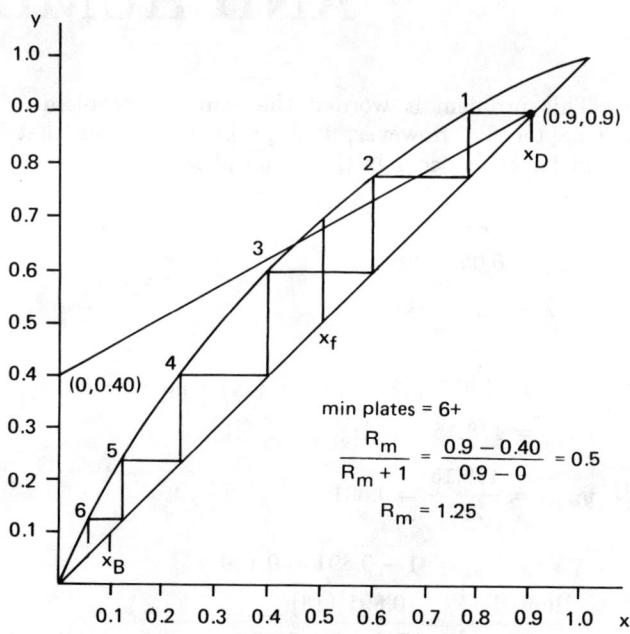

5. The sample is a point on the operating line.

(a) The operating line passes through (0.52, 0.61) and (0.95, 0.95).

$$\frac{L}{V} = \frac{R}{R+1}$$
$$= \frac{0.95 - 0.61}{0.94 - 0.52} = 0.791$$
$$R = \boxed{3.785}$$

(b) $\quad D = F\dfrac{x_f - x_B}{x_D - x_B}$
$$= (1000)\left(\frac{0.3 - 0.06}{0.95 - 0.06}\right)$$
$$= 269.7 \text{ lbmoles/hr}$$
$$V = D(R+1) = (269.7)(4.785)$$
$$= 1290.5 \text{ lbmoles/hr}$$

$$H_{v \text{ benzene}} = \left(94.14 \frac{\text{cal}}{\text{g}}\right)\left(78.1 \frac{\text{g}}{\text{gmol}}\right) = 7354$$

$$\Delta H_{v \text{ toluene}} = (86.8)(92.2) = 7999$$

$$\Delta H_{v \text{ mix}} = (7354)(0.95) + (7999)(0.05)$$
$$= 7386 \text{ cal/gmol}$$
$$= (7386)(1.8)$$
$$= 13{,}295 \text{ BTU/lbmole}$$

$$Q = (1290.5)(13{,}295)$$
$$= 1.716 \times 10^7 \text{ BTU/hr}$$

Assume cooling water.

in: 80°F   out: 120°F

Assume that the boiling point is 180°F.

$$\Delta t_1 = 180 - 80 = 100°F$$
$$\Delta t_2 = 180 - 120 = 60°F$$
$$\Delta t_{avg} = 80°F$$

Assume $U = 300$ BTU/hr-ft^2-°F.

$$Q = UA\Delta t$$
$$1.716 \times 10^7 = 300(A)80$$
$$A = \boxed{716 \text{ ft}^2}$$

(c) A preheater can be used depending upon whether the reflux ratio, total vapor, and reboiler rates are held constant. Preheating the feed would increase the vaporate in the column, improving the production rate while putting a higher load on the reboiler and condenser.

6. Normal hexane and n-heptane form an almost ideal solution and fit the criteria needed to apply the McCabe-Thiele method of equilibrium stage calculations. Thus, an x-y diagram is sufficient to evaluate the column operation.

(a) Reducing the reflux ratio from infinity (total) reflux to 10 will cause the distillate and bottoms composition to move slightly toward the feed composition. This happens because there is less room on the x-y diagram to fit the fixed number of equilibrium stages in the column.

(b) Reducing the reflux ratio to a value of 1 will give a slope of 0.5 to the upper operating line. This will intersect the equilibrium line if it starts from the total reflux composition of $x_D = 0.98$. The column will operate at this reflux ratio, but the distillate composition will be greatly reduced to a lower value, perhaps $x_D = 0.7$, and the bottoms will contain more n-hexane.

(c) Increasing the steam pressure will increase the boil-up rate and tend to increase the reboiler reflux ratio. Consequently, the overhead composition, x_D, will increase, and the bottoms, x_B, will decrease (an increase in separation of the higher temperature boiling component). The increased vapor rate may cause flooding or partial flooding of the trays and cause a lower tray efficiency, decreasing the overhead composition.

(d) Increasing the liquid and vapor rates can be accomplished by increasing the feed rate, reboiler heat input, and condenser cooling rate. The latter two will increase overhead and decrease bottoms composition. Increasing the feed rate will increase vapor and liquid rates and cause flooding. The result will be a lowering of plate efficiency, negating the improvements gained by changing reboiler heat input and condenser cooling water rate.

7.

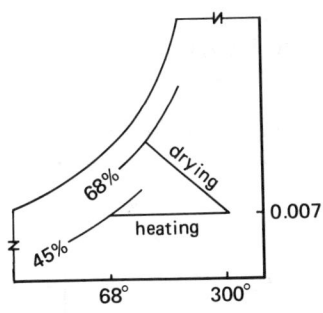

dry material = $(0.97)(30,000)$
$\qquad\qquad\quad = 29,100$ lbm/day

$$\text{feed} = \frac{29,100}{0.67}$$
$$= 43,433 \text{ lbm/day}$$

water removed = $43,433 - 30,000$
$\qquad\qquad\qquad = 13,433$ lbm/day

(a) From the chart,

inlet humidity = 0.007 lbm H_2O/lbm dry air
heated air wet bulb = 105°F = outlet wet bulb
outlet temp = 116°F
outlet humidity = 0.048 lbm H_2O/lbm air

$$\text{dry air} = \frac{\frac{13{,}433}{24}}{0.048 - 0.007}$$

$$= 13{,}659 \text{ lbm/hr}$$

The specific volume of air has a relative humidity of 45% and a temperature of $68°F$.

$$\text{air input} = \left(13.4 \, \frac{\text{ft}^3}{\text{lbm air}}\right)\left(13{,}659 \, \frac{\text{lbm air}}{\text{hr}}\right)\left(\frac{1}{60}\right)$$

$$= \boxed{3051 \text{ cfm}}$$

(b) humid heat $= 0.24 + 0.45H$
$= 0.24 + (0.45)(0.007)$
$= 0.243 \text{ BTU/°F lbm air}$

$$Q = \left(13{,}659 \, \frac{\text{lbm air}}{\text{hr}}\right)\left(0.243 \, \frac{\text{BTU}}{\text{°F-lbm}}\right)(300 - 68)$$

$$= \boxed{770{,}040 \text{ BTU/hr}}$$

8.

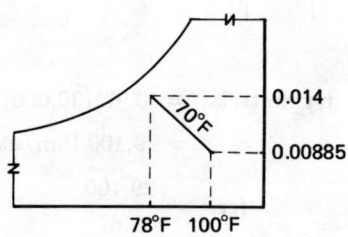

From the psychrometric chart,

(a) The humidity of the air to the tower is 0.00885 lbm H_2O/lbm air.

(b) The temperature of the humidified air is $78°F$.

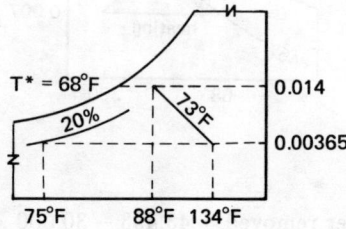

From the psychrometric chart,

(c) The water temperature is $73°F$.

(d) The temperature of the air leaving the heater is $134°F$.

(e) The temperature of the humidified air is $88°F$.

H_{air} from preheater: 29.5 BTU/lbm
H_{air} to preheater: 14.5 BTU/lbm

(f) The amount of heat added is

$$29.5 - 14.5 = \boxed{15 \text{ BTU/lbm}}$$

9. The initial free moisture is

$$0.3 - 0.02 = 0.28 \text{ lbm } H_2O/\text{dry solid}$$

The final free moisture is

$$0.1 - 0.02 = 0.08 \text{ lbm } H_2O/\text{dry solid}$$

The drying time needed to reduce the moisture content to 10 lbm water/lbm dry solid is 6 hours.

$$K(0.28 - 0.16) + K(0.16)\ln\left(\frac{0.16}{0.08}\right)$$

$$K = 25.97$$

$$\theta_r = K(0.12) + K\ln\left(\frac{0.16}{0.04}\right)$$

$$= \boxed{8.86 \text{ hours}}$$

10.

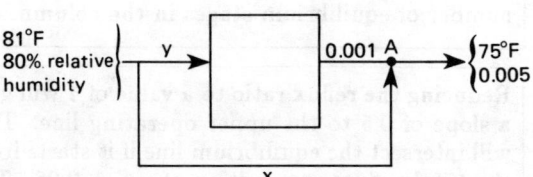

Use a basis of 100 lbm dry air.

From the psychrometric chart, the air entering is 0.018 lbm H_2O/lbm dry air, and the lbm dry air bypassed $= x$.

The H_2O balance around point A is

$$y(0.001) + x(0.018) = (100)(0.005)$$

$$y + x = 100$$

$$x = 23.53 \text{ lbm dry air}$$

$$\text{percent bypassed} = \boxed{23.53\%}$$

11.

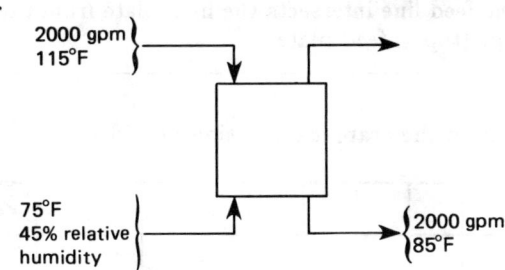

Assume that the air leaves saturated at 105°F.

$$H_{in} = 0.005 \text{ lbm } H_2O/\text{lbm dry air}$$
$$H_{out} = 0.05 \text{ lbm } H_2O/\text{lbm dry air}$$
$$\begin{aligned}q_{water} &= wc_p \Delta t \\ &= (2000)(8.33)(1)(115-85) \\ &= 5 \times 10^5 \text{ BTU/min}\end{aligned}$$

Determine enthalpy from the tables.

in:
$$\begin{aligned}h_{air} &= \frac{(h_{174°} + h_{76°})}{2} \\ &= \frac{17.778 + (0.45)(19.88) + 18.259 + (0.45)(21.31)}{2} \\ &= 27.286 \text{ BTU/dry air}\end{aligned}$$

out:
$$\begin{aligned}h_{air} &= \frac{(h_{104°} + h_{106°})}{2} \quad \text{(saturated)} \\ &= \frac{79.31 + 83.42}{2} \\ &= 81.37 \text{ BTU/lbm dry air}\end{aligned}$$

$$\begin{aligned}\Delta h_{tower} &= 81.37 - 27.29 \\ &= 54.08 \text{ BTU/lbm dry air}\end{aligned}$$

$$\frac{\text{lbm dry air}}{\text{min}} = \frac{5 \times 10^5}{54.08} = 9250 \text{ lbm dry air/min}$$

The specific volume of air at 75°F and 45% relative humidity is

$$\frac{13.398 + (0.45)(0.364) + 13.449 + (0.45)(0.422)}{2}$$
$$= 13.6 \text{ ft}^3/\text{lbm dry air}$$

(a) $\quad$ volume $= (13.6)(9250)$
$$= \boxed{125{,}803 \text{ ft}^3/\text{min}}$$

(b) make up $H_2O = \dfrac{(1.9)(0.05 - 0.005)(9250)}{8.33}$
$$= \boxed{94.9 \text{ gpm}}$$

(c) $\boxed{\text{At the boiling point of an azeotrope, the vapor and liquid composition are the same.}}$

12. The air properties at 29.92 in Hg are

$$H_{in} = 0.0035 \text{ lbm } H_2O/\text{lbm dry air (w.b.} = 30°F)$$
$$H_{out} = 0.3289 \text{ lbm } H_2O/\text{lbm dry air (w.b.} = 162.5°F)$$

To correct for pressure, use the tables in Perry's *Chemical Engineers' Handbook*.

in:
$$\begin{aligned}\Delta H &= 0.6 - \left(\frac{70}{24}\right)(0.01)(0.6) \\ &= 0.582 \text{ grains } H_2O/\text{lbm dry air} \\ &= 0.000083 \text{ lbm } H_2O/\text{lbm dry air} \\ H &= 0.0034 \text{ lbm } H_2O/\text{lbm dry air (at 14.3 psia)}\end{aligned}$$

out:
$$\begin{aligned}\Delta H &= 56 - \left(\frac{180}{24}\right)(0.1)(56) \\ &= 51.8 \text{ grains/lbm} \\ &= 0.0074 \text{ lbm/lbm} \\ H &= 0.3289 - 0.0074 \\ &= 0.3215 \text{ lbm } H_2O/\text{lbm dry air}\end{aligned}$$

evaporated:

$$0.3215 - 0.0034 = 0.3181 \text{ lbm } H_2O/\text{lbm dry air}$$

$$\begin{aligned}\text{lbm dry air} &= \left(\frac{1}{0.3181}\right)\left(12 \frac{\text{lbm } H_2O}{\text{hr}}\right) \\ &= 37.72 \text{ lbm dry air/hr}\end{aligned}$$

$$\begin{aligned}\text{humid volume} &= \left[\frac{(0.754)(t+459.8)}{29.1}\right] \\ &\quad [1 + (0.0034)(1.606)] \\ &= 13.80 \text{ ft}^3/\text{lbm dry air}\end{aligned}$$

$$\begin{aligned}Q &= \frac{(37.72)(13.80)}{60} \\ &= \boxed{8.67 \text{ ft}^3/\text{min}}\end{aligned}$$

13. Refer to the graphical solution provided.

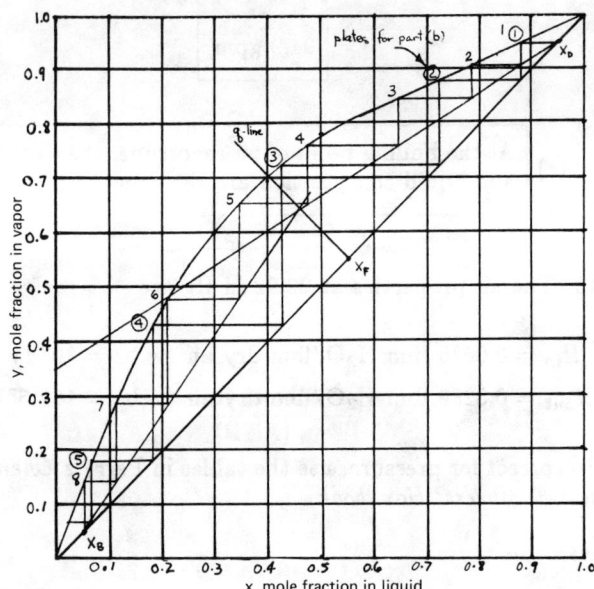

$$R_{D,\,min} = \frac{x_D - y'}{y' - x'}$$

x', y' are the coordinates of the intersection of the feed line and equilibrium line.

$$R_{D,\,min} = \frac{0.95 - 0.70}{0.70 - 0.40} = \boxed{0.833}$$

Refer to the graphical solution provided.

(a) For total reflux use $\boxed{4.6 \text{ plates } + \text{ reboiler.}}$

The reflux ratio is

$$2(R_{D\,min}) = 1.667$$

The y-intercept in the rectifying section is obtained when $x = 0$ for the equation

$$y = \frac{R_D}{R_D + 1} x + \frac{x_D}{R_D + 1}$$

$$y' = \frac{x_D}{R_D + 1} = \frac{0.95}{2.667} = 0.356$$

The strip operating line is drawn from the intersection of the rectifying operating line with the q line at $x_B = 0.05$.

(b) $\boxed{\text{The number of plates is 7.7 + reboiler.}}$

(c) $\boxed{\text{The number of actual plates is } 7.7/0.6 = 12.8 + \text{reboiler.}}$

(d) $\boxed{\text{The feed line intersects the fifth plate from the top. (top = feed plate)}}$

14. Refer to the graphical solution provided.

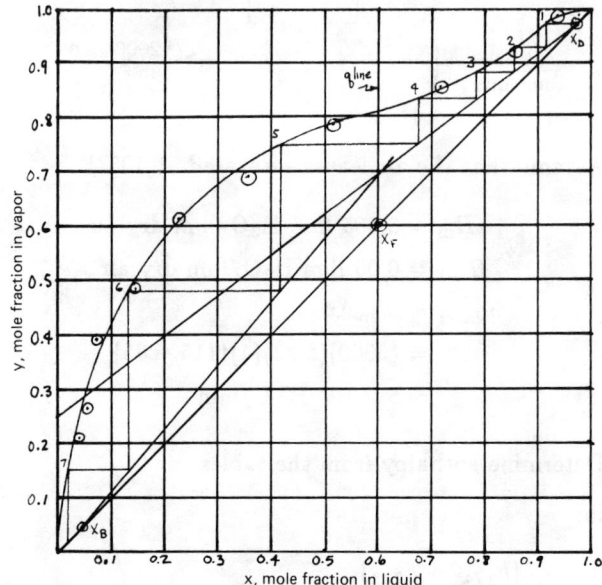

$$y = \frac{R}{R+1} x + \frac{x_D}{R+1}$$

At $x = 0$,

$$y = \frac{0.97}{3+1} = 0.2425$$

The operating line is constructed from (0.97, 0.97) to (0, 0.2425).

The feed line (q line) is the vertical strip operating line drawn from the q line intersection to (0.04, 0.04).

$$\text{theoretical plates } = 6.8 - 1 = 5.8 + \text{reboiler}$$
$$\text{actual plates } = \frac{5.8}{0.65} = 8.9 + \text{reboiler}$$

(a) $\boxed{\text{Use 12 plates for a safety factor.}}$

(b) $\boxed{\text{The feed plate on the theoretical 4.2 plate is the seventh actual from the top.}}$

SOLUTIONS FOR CHAPTER 10
LIQUID-LIQUID AND SOLID-LIQUID PROCESSES

1. Extract no. 1

	start	add	after mix	in soln	dry solids	drain soln	drain solids
sand	200	–	200	–	200	–	200
KCl	800	50	850	447[a]	403	193[b]	596
H$_2$O	–	950	950	950	–	410	410
$\sum$	1000	1000	2000	1397	603	603	1206

[a] H$_2$O can hold $(950)(0.471) = 447$ lbm KCl

[b] KCl $= \left(\dfrac{603}{1397}\right)(447) = 193$ lbm

Extract no. 2

	start	add	after mix	in soln	dry solids	drain soln	drain solids
sand	200		200	–	200	–	200
KCl	596		596	596	–[a]	53.9[b]	53.9
H$_2$O	410	1206	1616	1616	–	146.1	146.1
$\sum$	1206	1206	2412	2212	200	200	400

dry	solids	wgt %
sand	200	78.8
KCl	53.9	21.2
H$_2$O	–	
$\sum$	253.9	100

[a] H$_2$O in extract no. 2 can hold $(1616)(0.471) = 761$ lbm KCl. All dissolves.

[b] KCl $= \left(\dfrac{200}{2212}\right)(596) = 53.9$ lbm

2.

extract	extract no. 1	extract no. 2	$\sum$	wgt%
KCl	254	542.1	796.1	28.4
H$_2$O	540	1469.9	2009.9	71.6
$\sum$	794	2012.0	2806.0	100

3. If N is the amount of nicotine extracted into kerosene,

$$\dfrac{N \text{ lbm nicotine}}{150 \text{ lbm kerosene}} = (0.91)\left(\dfrac{1-N \text{ lbm nicotine}}{99 \text{ lbm H}_2\text{O}}\right)$$

$$N = 0.5796$$

$$\text{percent extracted} = \left(\dfrac{0.5796}{1}\right)(100)$$

$$= \boxed{57.96\%}$$

4.

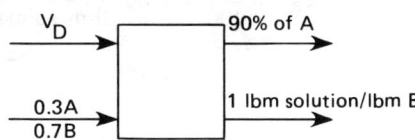

basis: lbm A + B

$$V_o = \dfrac{\text{lbm solvent feed}}{\text{lbm (A+B)}}$$

$$Y_1 = \dfrac{0.3}{V_o + 0.3}$$

90% of A in extract $= (0.9)(0.3) = 0.27$ lbm

A in underflow $= 0.3 - 0.27 = 0.03$ lbm

The amount of solution in the underflow is

$$(0.7 \text{ lbm B})\left(\dfrac{\text{lbm solution}}{\text{lbm B}}\right) = 0.7 \text{ lbm solution}$$

solvent in underflow $= 0.7 - 0.03 = 0.67$ lbm

$$Y_1 = \dfrac{0.03}{0.7} = 0.0428 \dfrac{\text{lbm A}}{\text{lbm solution}} \text{ overflow}$$

$$= \boxed{0.0428}$$

$$= \dfrac{0.3}{V_o + 0.3} = 6.7 \dfrac{\text{lbm solvent}}{\text{lbm(A+B)}}$$

5.

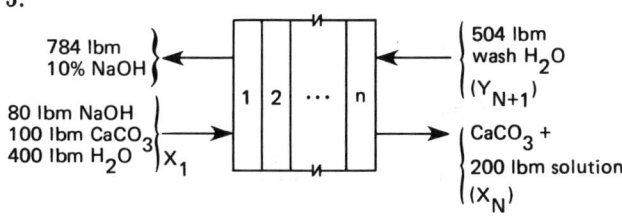

The NaOH balance around stage 1 is

$$80 + W = 78.4 + (0.1)(200)$$

$$W = 18.4 \text{ lbm in wash entering stage 1}$$

The wash entering stage 1 is

$$y_2 = \dfrac{18.4}{504} = 0.0365 \dfrac{\text{lbm NaOH}}{\text{lbm solution}}$$

The overall NaOH balance is

$$80 + 0 = 78.4 + W_{\text{in underflow}}$$

$$W_{\text{in underflow}} = 1.6$$

$$X_N = \frac{1.6}{200} = 0.008 \, \frac{\text{lbm NaOH}}{\text{lbm solution}}$$

$$X_o = \frac{80}{480} = 0.1667 \, \frac{\text{lbm NaOH}}{\text{lbm solution}}$$

$$X_1 = 0.1$$

$$n - 1 = \frac{\ln\left(\frac{y_{N+1} - X_n}{y_2 - X_1}\right)}{\ln\left(\frac{y_{N+1} - y_2}{X_n - X_1}\right)} = \frac{\ln\left(\frac{0 - 0.008}{0.0365 - 0.1}\right)}{\ln\left(\frac{0 - 0.0365}{0.008 - 0.1}\right)}$$

$$= 2.24$$

$$n = \boxed{3.24 \text{ stages}}$$

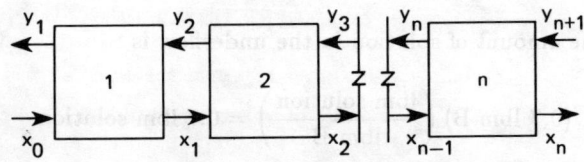

If the underflow is 1 lbm solution/lbm $CaCO_3$, then $Y_1 = 0.1$.

$$\text{underflow} = 100 \text{ lbm } CaCO_3$$
$$+ (1)(100 \text{ lbm } CaCO_3 \text{ in solution})$$
$$= 200 \text{ lbm}$$

$$\text{wash water} = 200 + 784 - 580 = 404 \text{ lbm}$$

$$X_N = \frac{1.6}{100} = 0.016 \text{ lbm/lbm solution}$$

The balance around stage 1 is

$$y_1(784) + X_1(100) = 80 + y_2(404)$$

$$y_1 = 0.1$$
$$x_1 = 0.1$$

$$78.4 + 10 = 80 + y_2(404)$$
$$y_2 = 0.0208$$

$$n - 1 = \frac{\ln\left(\frac{0 - 0.016}{0.0208 - 0.1}\right)}{\ln\left(\frac{0 - 0.0208}{0.016 - 0.1}\right)} = 1.14 \text{ stages}$$

$$n = \boxed{2.14 \text{ stages}}$$

6.
$$\frac{1}{U_1} = \frac{1}{500} = 0.002 \text{ hr-ft}^2\text{-}°F/BTU$$

$$\frac{1}{U_2} = \frac{1}{400} = 0.0025 \text{ hr-ft}^2\text{-}°F/BTU$$

$$\frac{1}{U_3} = \frac{1}{300} = 0.0033 \text{ hr-ft}^2\text{-}°F/BTU$$

$$\sum \frac{1}{U} = 0.0078 \text{ hr-ft}^2\text{-}°F/BTU$$

(a) $$q \left(\sum \frac{1}{U} \right) = A \sum \Delta t_c$$

$$\sum \Delta t_c = 260 - 95 - (12 + 6 + 3) = 144°F$$

$$q = (20{,}000)(939)(0.93)$$
$$= 17{,}400{,}000 \text{ BTU/hr}$$

$$A = \frac{q \left(\sum \frac{1}{U} \right)}{\sum \Delta t_c} = \frac{(1.74 \times 10^7)(7.8 \times 10^{-3})}{144}$$

$$= \boxed{940 \text{ ft}^2/\text{effect}}$$

(b) $$\Delta t_1 = \left(\frac{\frac{1}{U_1}}{\sum \frac{1}{U}} \right) \sum \Delta t_c$$

$$= \left(\frac{0.002}{0.0078} \right)(144) = 37°F$$

$$\Delta t_2 = \left(\frac{0.0025}{0.0078} \right)(144) = 46°F$$

$$\Delta t_3 = \left(\frac{0.0033}{0.0078} \right)(144) = 61°F$$

The temperatures are

chest I = $\boxed{260°F}$

vapor I = chest II = $223 - 12 = \boxed{211°F}$

body I = $260 - 37 = \boxed{223°F}$

chest III = vapor II = $165 - 6 = \boxed{159°F}$

body II = $211 - 46 = \boxed{165°F}$

vapor III = condenser = $98 - 3 = \boxed{95°F}$

body III = $159 - 61 = \boxed{98°F}$

(c) The vapor from each effect is approximately 20,000 lbm/hr.

$$(0.93)(60,000) = 0.6w_f \left(\frac{94}{6} - \frac{50}{50}\right)$$

$$w_f = \frac{(0.93)(60,000)}{(0.06)(14.8)} = \boxed{67,500 \text{ lbm/hr}}$$

7.

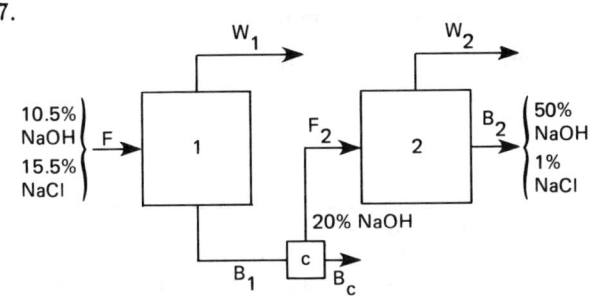

$$(100 \text{ tons/day})\left(\frac{1}{22}\right) = 0.5 B_2$$

$$B_2 = 18,182 \text{ lbm/hr}$$

The balance around effect 2 is

$$\text{NaOH} \quad 0.2 F_2 = 0.5 B_2$$
$$\text{NaCl} \quad X F_2 = 0.01 B_2$$

$$X = 0.004 = 0.4 \text{ wgt \% in mother liquor}$$

The composition of B_c is

$$\left(\frac{5 \text{ lbm H}_2\text{O}}{100 \text{ lbm mixture } B_c}\right)\left(\frac{0.2 \text{ lbm NaOH}}{0.796 \text{ lbm H}_2\text{O}}\right)$$

$$= \frac{1.256 \text{ lbm NaOH}}{100 \text{ lbm mixture } B_c}$$

B_c 5% H_2O, 1.256% NaOH, 93.74% NaCl
B_2 NaOH: $(0.5)(18,182) = 9091$ lbm/hr
  NaCl: $(0.01)(18,182) = 181.8$ lbm/hr
  H_2O: $-(9091 + 181.8) + 18,182 = 8909.2$ lbm/hr
F_2 NaCl: $0.004 F_2 = 181.8$ lbm/hr

$$F_2 = 45,455 \text{ lbm/hr}$$

The overall balance is

$$F = W_1 + W_2 + B_c + B_2$$

The overall NaOH balance is

$$0.105 F = (0.5)(18,182) + 0.01256 B_c$$

The overall NaCl balance is

$$0.155 F = (0.01)(18,182) + 0.9374 B_c$$

$$F = 88,304 \text{ lbm/hr}$$
$$B_c = 14,407 \text{ lbm/hr}$$

The material balance is in lbm/hr.

	NaOH	NaCl	H_2O	Σ
F	9271	13,687	65,346	88,304
B_1	9271	13,687	36,902	59,860
W_1	–	–	28,444	28,444
B_c	182	13,505	720	14,407
F_2	9091	181.8	36,182	45,455
W_2	–	–	27,273	27,273
B_2	9091	181.8	8909	18,182

8. First, perform a heat balance around the evaporator.

evaporation: $(14,000)(970) = 1.36 \times 10^7$ BTU/hr

heating: $(8.33)(60)(100)$
$\times (1.05)(1)(212 - 185) = 1.42 \times 10^6$ BTU/hr

heat required $= 1.502 \times 10^7$ BTU/hr

Let $x =$ lbm/hr of 120 psia steam
$=$ lbm/hr of vapor compressed from 212°F to 233°F

$H_{\text{in}} = 1190.4x$ 120 psi steam
$H = 1150.4x$ 212°F vapor
$h_{\text{condensate}} = (2x)(201.26)$
$H_{\text{balance}} = 1190.4x + (1150.4x - 2x)(201.26)$
$= 1.502 \times 10^7$

$x = \boxed{7750 \text{ lbm/hr steam}}$

Other types of recompression are mechanical and centrifugal.

9. $\mu = (0.982)(6.72 \times 10^{-4}) = 6.6 \times 10^{-4}$ lbm/ft-sec

$$K = D_p \left[\frac{g(\rho_p - \rho)}{\mu^2}\right]^{\frac{1}{3}}$$

Stokes' law: $V_t = \dfrac{gD_p^2(\rho_p - \rho)}{18\mu}$

intermediate law: $V_t = \dfrac{0.153 g^{0.71} D_p^{1.14}(\rho_p - \rho)^{0.71}}{\rho^{0.29} \mu^{0.43}}$

For galena,

$$K = D_p \left[\frac{(32.17)(62.3)(6.5)(62.4)}{(6.6 \times 10^{-4})^2}\right]^{\frac{1}{3}}$$

$$= D_p (1.231 \times 10^4)$$

Since $D_p = 0.01$, $K = 10.5$ and obeys the intermediate law.

$$v_t = \frac{(0.153)(32.17)^{0.71}\left(\frac{0.01}{12}\right)^{1.14}[(6.5)(62.4)]^{0.71}}{(62.3)^{0.29}(6.6\times 10^{-4})^{0.43}}$$

$$= 0.279 \text{ ft/sec}$$

$$t = \frac{5}{0.279} = \boxed{18 \text{ sec}}$$

0.001 in diameter $K = 1.025 < 3.5$ (Stokes)

$$v_t = \frac{(32.17)(0.833)^2(10)^{-8}(406)}{(18)(6.6\times 10^{-4})}$$

$$= 0.00761 \text{ ft/sec}$$

$$t = \frac{5}{0.00761} = \boxed{657 \text{ sec}}$$

Similarly, for quartz, $D_p = 0.01$, $K = 6.5$ and obeys the intermediate law.

$$v_t = 0.1049 \text{ ft/sec}$$

$$t = \boxed{48 \text{ sec}}$$

$$D_p = 0.001 \text{ (Stokes)}$$

$$v_t = 0.001932 \text{ ft/sec}$$

$$t = \boxed{43 \text{ min}}$$

10. $$Q_c = \frac{(\rho_s - \rho)D_p^2 V \omega^2 r}{9\mu s}$$

$$V = \left(\frac{16}{12}\right)\pi\frac{(D_1^2 - D_2^2)}{4} = \left(\frac{\pi}{3}\right)[(2)^2 - (1.5)^2]$$

$$= 1.83 \text{ ft}^3$$

$$\mu = 3 \text{ cp} = 2.016\times 10^{-3} \text{ lbm/ft-sec}$$

$$(\rho_s - \rho) = (1.6 - 1.3)(62.4) = 18.72 \text{ lbm/ft}^3$$

$$\omega = \frac{(1000)(2\pi)}{60} = 104.7 \text{ sec}^{-1}$$

$$r = 1 \text{ ft}$$

$$s = 0.25 \text{ ft}$$

$$D_p = 30\,\mu m = 9.84\times 10^{-5} \text{ ft}$$

$$Q_c = \frac{(18.72)(9.84\times 10^{-5})^2(1.83)(104.7)^2(1)}{(9)(2.016\times 10^{-3})(0.25)}$$

$$= 0.8016 \text{ ft}^3/\text{sec}$$

$$= \boxed{359.8 \text{ gpm}}$$

11. Assume:
- cake mechanism
- cake incompressible
- clinical filters run at same temperature
- clinical uses same filter
- clinical minimum flow rate is 0.1 ml/sec or 6 ml/min
- clinical maximum volume is 1 ℓ
- clinical filter area is 9 in^2

t	V	Δt	ΔV	$\frac{\Delta t}{\Delta V}$	V	from plot $\frac{\Delta t}{\Delta V}$
0.0	0	–	–	–	–	–
2.0	27	2.0	27	0.0741	13.5	0.0685
10.3	66	8.3	39	0.2128	46.5	0.2246
14.5	78	4.2	12	0.3500	72.0	0.3453
23.9	100	9.4	22	0.4273	89.0	0.4257

From the plot,

$$\frac{\Delta t}{\Delta V} = 0.004732 V + 0.004597$$

For constant pressure filtration,

$$\frac{\Delta t}{\Delta V} = K_p V + B$$

$$K_p = 0.004732 = \frac{w\alpha\mu}{A^2 \Delta P}$$

$$B = 0.004597 = \frac{r\mu}{A\Delta P}$$

$$A = 3 \text{ in}^2$$

$$\Delta P = 2 \text{ atm}$$

Therefore,

$$w\alpha\mu = (0.004732)(3)^2(2)$$

$$= 0.08518 \text{ min-in}^4\text{-atm/ml}^2$$

$$r\mu = (0.004597)(3)(2)$$

$$= 0.02759 \text{ min-in}^2\text{-atm/ml}$$

For constant rate filtration,

$$\Delta P = K_r V + C$$

$$K_r = \frac{w\alpha\mu Q}{A^2}$$

$$C = \frac{r\mu Q}{A}$$

Therefore,

$$K_r = \frac{(0.08518)(6)}{(9)^2} = 0.006309 \text{ atm/ml}$$

$$C = \frac{(0.02759)(6)}{9} = 0.018391 \text{ atm}$$

When $V = 1000$ ml,

$$\Delta P = (0.006309)(1000) + 0.018391$$
$$= \boxed{6.328 \text{ atm}}$$

cake mechanism

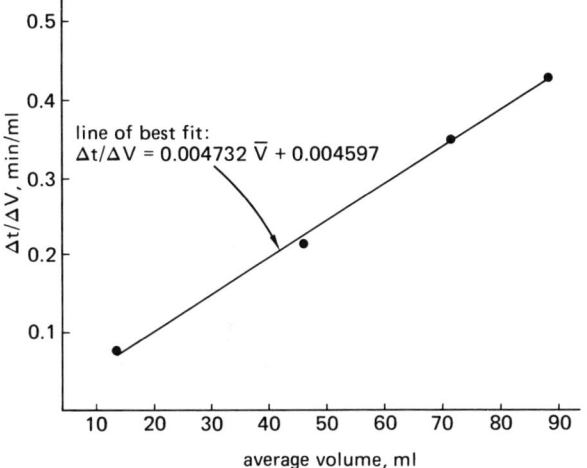

line of best fit:
$\Delta t/\Delta V = 0.004732 \bar{V} + 0.004597$

12. $\text{rate} = \sqrt{\dfrac{2\Delta pf}{\mu\alpha\omega\theta_\pi}}$

Assume that at the new rate Δp, f, and properties of the cake are constant.

$$\frac{r_2}{r_1} = \sqrt{\frac{\theta_{\pi_1}}{\theta_{\pi_2}}}$$

Doubling the rpm halves the cycle time.

$$\frac{r_2}{r_1} = \sqrt{\frac{1}{\frac{1}{2}}} = 1.414$$

$$r_2 = \boxed{r_1(1.414) \text{ or } 41.4\% \text{ increase}}$$

SOLUTIONS FOR CHAPTER 11
KINETICS

1. (a) $\ln\left(\frac{10}{1}\right) = \left(\frac{E}{R}\right)\left(\frac{1}{400} - \frac{1}{500}\right)$

    $R = 1.987 \text{ cal/gmol·K}$

    $\frac{E}{R} = 4605.1 \text{ K}$

    $E = \boxed{9.15 \text{ kcal/gmol}}$

   (b) $\ln\left(\frac{k_2}{k_1}\right) = \left(\frac{E}{R}\right)\left(\frac{1}{400} - \frac{1}{450}\right)$

    $= (4605.1)\left(\frac{1}{400} - \frac{1}{450}\right)$

    $= 1.27919$

    $\frac{k_2}{k_1} = \boxed{3.59 \text{ times as fast at 450K}}$

2. $\frac{dC_A}{dt} = kC_A^{0.5}$

    $C_A = C_{A_o}(1 - X_A)$

    $kt = 2C_{A_o}^{0.5}\left[1 - (1 - X_A)^{0.5}\right]$

   At $t = 10$,

    $X_A = 0.75$

    $k = \frac{C_{A_o}^{0.5}}{10}$

   At $t = 30$,

    $\left(\frac{C_{A_o}^{\frac{1}{2}}}{10}\right)(30) = C_{A_o}^{\frac{1}{2}}\left[1 - (1 - X_A)^{0.5}\right]$

    $X_A = 0.75 \quad \text{(not possible)}$

   Since at $t = 20$,

    $\left(\frac{C_{A_o}^{\frac{1}{2}}}{10}\right)(20) = 2C_{A_o}^{\frac{1}{2}}\left[1 - (1 - X_A)^{\frac{1}{2}}\right]$

    $X_A = 1.00$

   Therefore, at $t = 30$,

    $\boxed{X_A = 1.00 \text{ also}}$

3. $\ln\left(\frac{1}{1 - X_A}\right) = kt$

    $\ln\left(\frac{1}{1 - 0.5}\right) = \ln 2 = k(5)$

    $k = \frac{\ln 2}{5}$

   At $X_A = 0.75$,

    $\ln\left(\frac{1}{1 - 0.75}\right) = \ln 4 = kt$

    $= \frac{\ln 2}{5} t$

    $t = (5)\left(\frac{\ln 4}{\ln 2}\right)$

    $= 10 \text{ minutes}$

   additional time $= 10 - 5 = \boxed{5 \text{ minutes}}$

4. For an ideal gas,

    $\frac{p_1}{T_1} = \frac{p_2}{T_2}$

    $p_2 = (1)\left(\frac{373}{298}\right) = 1.25 \text{ atm}$

    $\frac{-dC_A}{dt} = kC_A^2$

    $C_A = \frac{p_A}{RT}$

    $RT\left(\frac{1}{p_A} - \frac{1}{1.25}\right) = kt$

    $R = 0.0821 \text{ ℓ·atm/mol·K}$

    $2A \longrightarrow B$

    $\Delta n = 1 - 2 = -1$

    $p_A = p_{A_o} - \left(\frac{2}{-1}\right)(\pi - \pi_o)$

    $p_A = 2\pi - 1.25$

t	π	p_A	kt
0	1.25	1.25	0
1	1.14	1.03	5.23
2	1.04	0.83	12.40
3	0.982	0.714	18.39
4	0.940	0.63	24.11
5	0.905	0.56	30.19
6	0.870	0.49	38.00
7	0.850	0.45	43.55
8	0.832	0.414	49.47
9	0.815	0.38	56.09
10	0.800	0.35	63.00
15	0.754	0.258	94.20
20	0.728	0.206	124.16
$\Sigma = 90$			$\Sigma = 558.79$

$$k = \frac{558.79}{90} = 6.2088$$

The rate equation at 373K is

$$\boxed{\frac{1}{p_A} = 0.203t + 0.8}$$

p_A is the partial pressure of A in atm, and t is time, in minutes.

5. The basis is 1 lbmole chlorohydrin.

	lbmoles	lbm	
chlorohydrin	1.0	80.52	feed
water		241.5	414.44
bicarbonate	1.10	92.42	
glycol	0.99		
NaCl	0.99		product
NaHCO$_3$	0.11		370.9 lbm
chlorohydrin	0.01		
water		241.5	
CO$_2$ ↑	0.99	43.6	offgas 43.6 lbm

$$V = \frac{370.9}{(8.33)(1.1)} = 40.4 \text{ gal}$$

$$X_A = 0.99$$

$$k_c = 1250 \text{ gal-unit charge/hr}$$

$$n_A = 0.01$$

$$n_B = 0.11$$

$$\frac{V_2}{F} = \frac{X_A}{k_c\left(\frac{n_A}{V}\right)\left(\frac{n_B}{V}\right)}$$

$$= \frac{0.99}{(1250)\left(\frac{0.01}{40.4}\right)\left(\frac{0.11}{40.4}\right)}$$

$$\frac{V_2}{F} = 1175 \text{ gal-hr/unit charge}$$

$$1 \text{ unit charge} = 61.5 \text{ lbm product}$$

$$F = \frac{10}{61.5}$$

$$= 0.1625 \text{ unit charge/hr}$$

$$V_2 = (0.1625)(1175)$$

$$= \boxed{191 \text{ gal}}$$

6. $(CH_3CO)_2O + H_2O \longrightarrow 2CH_3COOH$

$$\left(\frac{1}{V}\right)\left(\frac{dNa}{dt}\right) = -kC_{H_2O}C_A$$

(a) Since the solution is dilute, the H$_2$O concentration is constant, therefore the kC_{H_2O} is constant.

(b) For the integrated second-order batch,

$$\ln\left(\frac{1-X_B}{1-X_A}\right) = (C_{B_o} - C_{A_o})kt$$

$$C_B \approx C_{B_o}$$

$$X_B \approx 0$$

$$C_{B_o} \gg C_{A_o}$$

Therefore,

$$-\ln(1 - X_A) = C_{B_o}kt = k't$$

$$-\ln(1 - 0.7) = \boxed{0.0806t}$$

$$t = 14.9 \text{ min}$$

(c) $$k = k_o e^{-\frac{E}{RT}}$$

or $$kC_{H_2O} = k'_o e^{-\frac{E}{RT}}$$

$$T = 283.15 \text{K}$$

$$k'_o e^{\frac{-E}{R}} \text{ (at 283.15)} = 0.0567 \text{ }\ell/\text{mol·s}$$

$$T = 313.15 \text{K}$$

$$k'_o e^{\frac{-E}{R}} \text{ (at 313.15)} = 0.380 \text{ }\ell/\text{mol·s}$$

$$\left(\frac{E}{R}\right)\left(\frac{1}{283.15} - \frac{1}{313.15}\right) = 1.9$$

$$\frac{E}{R} = 5622 \text{K}$$

Therefore, at 313.15,

$$k'_o = 0.380 e^{\frac{5622}{313.15}}$$

$$k'_o = 2.38 \times 10^7$$

$$-r_A = \boxed{C_A(2.38 \times 10^7)e^{\frac{-5622}{T}}}$$

7. $\overset{(A)}{O_2} + \overset{(B)}{2NO} \longrightarrow 2NO_2$

(a) initially

	moles		moles at 90%
O$_2$	0.1053	$0.1053 - (0.9)\left(\frac{0.1842}{2}\right) =$	0.02241
NO	0.1842	$(0.1)(0.1842) =$	0.01842
N$_2$	0.7105		0.7105
NO$_2$	0	$(0.9)(0.1842) =$	0.16578
	1.0000		0.91711

The volume of 1 gmol of gas at 1 atm and 30°C is

$$V_R = (22.4)(1)\left(\frac{303.15}{273.15}\right)$$

$$= 24.86 \text{ }\ell$$

KINETICS

For the integrated third-order equation,

$$\frac{(2C_{A_o} - C_{B_o})(C_{B_o} - C_B)}{C_{B_o} C_B} + \ln\left(\frac{C_{A_o} C_B}{C_A C_{B_o}}\right)$$
$$= (2C_{A_o} - C_{B_o})^2 kt$$
$$C_i = \frac{n_i}{V}$$

The volume terms cancel the left side.

$$\frac{[(2)(0.1053) - 0.1842](0.1842 - 0.01842)}{(0.1842)(0.01842)}$$
$$+ \ln\left[\frac{(0.1053)(0.01842)}{(0.02241)(0.1842)}\right]$$
$$= \frac{[(2)(0.1053) - 0.1842]^2}{V^2} kt$$

$$t = \boxed{17.89 \text{ sec}}$$

(b)

i	x_i
O_2	$\frac{0.02241}{0.91711} = 0.0244$
NO	$\frac{0.01842}{0.91711} = 0.0201$
N_2	$\frac{0.7105}{0.91711} = 0.7747$
NO_2	$\frac{0.16578}{0.91711} = 0.1808$

The total pressure is proportional to the moles present.

$$p = (0.91711)(760) = \boxed{697 \text{ mm Hg}}$$

8. The total inlet feed = 9 gpm.

$$C_{A_o} = \frac{(3)(5)}{9} = 1.667 \text{ lbmole/gal}$$
$$C_{B_o} = \frac{(1.5)(4)}{9} = 0.667 \text{ lbmole/gal}$$

For an 80% conversion,

$$C_B = (0.2)(0.667) = 0.1334 \text{ lbmole/gal}$$
$$C_A = 1.667 - (0.667 - 0.1334) = 1.133 \text{ lbmole/gal}$$
$$\frac{V}{q} = \frac{0.667 - 0.1334}{(1)(1.133)(0.1344)} = 3.504 \text{ min}$$
$$q = 9 \text{ gpm}$$
$$V = (3.504)(9) = \boxed{31.53 \text{ gal}}$$

9. Using Arrhenius' plot,

$$\frac{d \ln k}{dT} = \frac{E}{RT^2}$$
$$k = k_o \left(2^{\frac{(T_2 - T_1)}{10}}\right)$$
$$\ln k = \ln k_o + \frac{T_2 - T_1}{10} \ln 2$$
$$\frac{d \ln k}{dT} = \frac{\ln 2}{10} = 0.06931 = \frac{E}{RT^2}$$
$$E = 1.383 \times 10^{-4} T^2$$

T, K	300	400	600	800	1000
E $\frac{\text{kcal}}{\text{gmol}}$	12.45	22.1	49.8	88.5	138.3

10. Assume a first-order reaction.

$$\ln\left(\frac{C_A}{C_{A_o}}\right) = -kt$$

(a)

100°C				120°C	150°C
C_A	$\frac{C_A}{C_{A_o}}$	$-\ln\left(\frac{C_A}{C_{A_o}}\right)$	t	t	t
15	1.0	0	0	0	0
12	0.8	0.2231	111	55.3	22.7
10.5	0.7	0.3567	178		
7.5	0.5	0.6931	346	172	70.5
6	0.4	0.9162	457		
4.5	0.3	1.2040	600	300	122.4
3.0	0.2	1.6094	804	402.2	164.1
1.5	0.1	2.3026	1145	573.1	232.0
		Σ=7.3051	Σ=3641		
		Σ=6.0322		Σ=1502.6	
		Σ=6.0322			Σ=611.7

Plot $-\ln\left(\frac{C_A}{C_{A_o}}\right)$ versus t. This yields a straight line.

$$\boxed{\text{Therefore, the reaction is first-order.}}$$

(b)
$$k_{100} = \frac{7.3051}{3641} = 2.006 \times 10^{-3} \text{ s}^{-1}$$
$$k_{120} = \frac{6.0322}{1502.6} = 4.0145 \times 10^{-3} \text{ s}^{-1}$$
$$k_{150} = \frac{6.0322}{611.7} = 9.861 \times 10^{-3} \text{ s}^{-1}$$

$$\ln\left(\frac{k_1}{k_2}\right) = -\frac{E}{R}\left(\frac{1}{T_1} - \frac{1}{T_2}\right)$$

$$-\frac{E}{R} = \frac{\ln\left(\frac{k_1}{k_2}\right)}{\frac{1}{T_1} - \frac{1}{T_2}}$$

$$\left(\frac{\ln\left(\frac{k_1}{k_2}\right)}{\frac{1}{T_1} - \frac{1}{T_2}}\right) = -5095\text{K}$$

$$\left(\frac{\ln\left(\frac{k_1}{k_3}\right)}{\frac{1}{T_1} - \frac{1}{T_3}}\right) = -5019\text{K}$$

$$\left(\frac{\ln\left(\frac{k_2}{k_3}\right)}{\frac{1}{T_2} - \frac{1}{T_3}}\right) = -4963\text{K}$$

$$\text{average} = 5026\text{K}$$

Since $k = k_o e^{-\frac{E}{RT}}$, at $T = 373$K,

$$k = 2.006 \times 10^{-3} \text{ s}^{-1}$$

$$-\frac{E}{R} = -5026\text{K}$$

$$k_o = ke^{\frac{E}{RT}} = 2.006 \times 10^{-3} e^{\frac{5026}{373}}$$

$$= 1.426 \times 10^3$$

$$\boxed{k = 1.426 \times 10^3 e^{\frac{-5026}{T}} \text{ s}^{-1}}$$

11. For the first-order reaction,

$$\ln\left(\frac{C_E}{C_{E_o}}\right) = -kt$$

$$k = \left(\frac{1}{t}\right)\ln\left(\frac{C_{E_o}}{C_E}\right)$$

$$k_{70} = \left(\frac{1}{10}\right)\ln\left(\frac{0.175}{0.175 - \frac{0.022}{2}}\right)$$

$$= 6.492 \times 10^{-3} \text{ min}^{-1}$$

$$k_{100} = \left(\frac{1}{10}\right)\ln\left(\frac{0.175}{0.175 - \frac{0.059}{2}}\right)$$

$$= 1.846 \times 10^{-2} \text{ min}^{-1}$$

$$\ln\left(\frac{k_{100}}{k_{70}}\right) = \frac{E}{R}\left(\frac{1}{530} - \frac{1}{560}\right) = 1.045$$

$$\frac{E}{R} = 10\,340\text{K}$$

At $170°$F $(630°$R$)$,

$$\ln\left(\frac{k_{170}}{k_{70}}\right) = (10\,340)\left(\frac{1}{530} - \frac{1}{630}\right)$$

$$= 3.097$$

$$k_{170} = 0.1436 \text{ min}^{-1}$$

For an 80% conversion,

$$\ln\left(\frac{E_o}{E}\right) = \ln\left(\frac{E_o}{0.2E_o}\right) = \ln 5$$

$$= 0.1436t$$

$$\boxed{t = 11.21 \text{ minutes}}$$

12. The reaction has a half-life of 1.3 hours.

$$\ln\left(\frac{1}{2}\right) = -k(1.3)$$

$$k = 0.532 \text{ hr}^{-1}$$

In batch,

$$\ln\left(\frac{A}{A_o}\right) = \ln 0.30 = -kt$$

$$t = 2.26 \text{ hr}$$

$$\frac{200 \text{ }\ell}{2.26 \text{ hr}} = 88.5 \text{ }\ell/\text{hr}$$

In the CSTR,

$$qA_o = qA + kVA$$

$$A = 0.3A_o$$

$$q = 0.3q + 0.3kV = \frac{0.3}{0.7}kV$$

$$= \left(\frac{0.3}{0.7}\right)(400)(0.532)$$

$$= 91.0 \text{ }\ell/\text{hr}$$

$$\boxed{\text{Choose the CSTR.}}$$

13. (a) $2P \longrightarrow Q + R$

$$\frac{1}{C_p} - \frac{1}{C_{p_o}} = kt = (0.2)(1)$$

$$\frac{1}{C_p} = 0.8666 \text{ }\ell/\text{mol}$$

$$C_p = 1.1538 \text{ mol}/\ell$$

Or, $1.5 - 1.1538 = 0.346$ mol/ℓ of P reacted. Therefore, the mol/ℓ of Q formed are

$$\frac{0.346}{2} = 0.173 \text{ mol}/\ell$$

The volume is

$$V = \frac{100}{0.173} = \boxed{578 \ \ell}$$

(b) Let q represent the volume feed rate.

$$qP_o - VkP^2 = qP$$

200 mol of P must disappear.

$$q(P - P_o) = 200$$
$$= (578)(0.2)(P)^2$$
$$P = 1.315 \text{ mol}/\ell$$

The batch reactor can be run as a continuous stirred tank reactor since P is less than 1.5.

$$q = \frac{200}{1.5 - 1.315} = \boxed{1081 \ \ell/\text{hr}}$$

14. $\tau = \dfrac{V}{v_o} = \dfrac{C_{A_o} X_A}{-r_A}$

$$C_{A_o} = C_{B_o}$$
$$\frac{dC}{dt} = kC_A C_B$$
$$= kC_{A_o}(1 - X_A)C_{B_o}(1 - X_B)$$
$$= kC_{A_o}^2 (1 - X_A)^2$$
$$-r_A = -r_B = r_o$$

$$\tau = \frac{C_{A_o} X_A}{-r_A} = \frac{2000}{6000}$$
$$= 0.33 \text{ hr} = 20 \text{ min}$$
$$= \frac{C_{A_o} X_A}{kC_{A_o}^2 (1 - X_A)^2}$$
$$= \frac{X_A}{(17.1)(0.2)(1 - X_A)}$$

Solving the quadratic equation,

$$X_A = 0.886$$
$$C_D = (0.886)(0.2)$$
$$= \boxed{0.177 \text{ lbmoles}/\text{ft}^3}$$

15. $2A \longrightarrow (2R + S)$

For a variable volume batch reactor,

$$tkC_{A_o} = \frac{(1 + \epsilon_A)X_A}{1 - X_A} + \epsilon_A \ln(1 - X_A)$$
$$\epsilon_A = \frac{3 - 2}{2} = \frac{1}{2}$$
$$C_{A_o} = \frac{n_{A_o}}{V_{A_o}} = \frac{P}{RT}$$
$$= \frac{1 \text{ atm}}{(0.0821)(273 + 950)}$$
$$= 9.959 \times 10^{-3} \text{ gmol}/\ell$$
$$k = 1200 \text{ cc/mol·s}$$
$$= 1.2 \ \ell/\text{mol·s}$$

At 90% decomposition,

$$X_A = 0.9$$
$$t(9.959 \times 10^{-3})(1.2) = \frac{\left(1 + \frac{1}{2}\right)(0.9)}{1 - 0.9} + \frac{1}{2}\ln(1 - 0.9)$$
$$t = \boxed{1033 \text{ s}}$$